Florian Leiß

Dendritic spines and structural plasticity in Drosophila

Florian Leiß

Dendritic spines and structural plasticity in Drosophila

Anatomical investigations of visual and olfactory interneurons in the fruit fly

Südwestdeutscher Verlag für Hochschulschriften

Impressum/Imprint (nur für Deutschland/ only for Germany)
Bibliografische Information der Deutschen Nationalbibliothek: Die Deutsche Nationalbibliothek verzeichnet diese Publikation in der Deutschen Nationalbibliografie; detaillierte bibliografische Daten sind im Internet über http://dnb.d-nb.de abrufbar.

Alle in diesem Buch genannten Marken und Produktnamen unterliegen warenzeichen-, marken- oder patentrechtlichem Schutz bzw. sind Warenzeichen oder eingetragene Warenzeichen der jeweiligen Inhaber. Die Wiedergabe von Marken, Produktnamen, Gebrauchsnamen, Handelsnamen, Warenbezeichnungen u.s.w. in diesem Werk berechtigt auch ohne besondere Kennzeichnung nicht zu der Annahme, dass solche Namen im Sinne der Warenzeichen- und Markenschutzgesetzgebung als frei zu betrachten wären und daher von jedermann benutzt werden dürften.

Verlag: Südwestdeutscher Verlag für Hochschulschriften Aktiengesellschaft & Co. KG
Dudweiler Landstr. 99, 66123 Saarbrücken, Deutschland
Telefon +49 681 37 20 271-1, Telefax +49 681 37 20 271-0
Email: info@svh-verlag.de
Zugl.: München, Ludwigs-Maximilians-Universität, Dissertation, 2009

Herstellung in Deutschland:
Schaltungsdienst Lange o.H.G., Berlin
Books on Demand GmbH, Norderstedt
Reha GmbH, Saarbrücken
Amazon Distribution GmbH, Leipzig
ISBN: 978-3-8381-0341-9

Imprint (only for USA, GB)
Bibliographic information published by the Deutsche Nationalbibliothek: The Deutsche Nationalbibliothek lists this publication in the Deutsche Nationalbibliografie; detailed bibliographic data are available in the Internet at http://dnb.d-nb.de.

Any brand names and product names mentioned in this book are subject to trademark, brand or patent protection and are trademarks or registered trademarks of their respective holders. The use of brand names, product names, common names, trade names, product descriptions etc. even without a particular marking in this works is in no way to be construed to mean that such names may be regarded as unrestricted in respect of trademark and brand protection legislation and could thus be used by anyone.

Publisher: Südwestdeutscher Verlag für Hochschulschriften Aktiengesellschaft & Co. KG
Dudweiler Landstr. 99, 66123 Saarbrücken, Germany
Phone +49 681 37 20 271-1, Fax +49 681 37 20 271-0
Email: info@svh-verlag.de

Printed in the U.S.A.
Printed in the U.K. by (see last page)
ISBN: 978-3-8381-0341-9

Copyright © 2010 by the author and Südwestdeutscher Verlag für Hochschulschriften Aktiengesellschaft & Co. KG and licensors
All rights reserved. Saarbrücken 2010

All figures are originally coloured. Please contact the author (florian.leiss@gmx.de) to obtain a digital document containing all figures in colour.

Table of Content

Index of figures	4
Abbreviations	5
Acknowledgements	7
Summary	9
Introduction	11
Material and Methods	25
Results	35
Discussion	75
Literature	97

Index of figures

Figure 1.1 | Schematic illustration of the olfactory circuit in *Drosophila* 19
Figure 3.1 | Different dendrite model systems 36
Figure 3.2 | LPTC overview 37
Figure 3.3 | *Drosophila* Lobula Plate Tangential Cells have spines 39
Figure 3.4 | Classification of dendritic spines of LPTCs 41
Figure 3.5 | LPTC spines receive synaptic input 42
Figure 3.6 | Dα7 is localized at dendritic spines 43
Figure 3.7 | Spine density is modulated by Rac1 45
Figure 3.8 | Mushroom body overview 48
Figure 3.9 | Actin-enriched microglomeruli in the mushroom body calyx 49
Figure 3.10 | Synaptic organization of calycal microglomeruli 51
Figure 3.11 | Acetylcholine receptors in Kenyon cells 53
Figure 3.12 | GABAergic interneurons and glial cells are present in the calyx 54
Figure 3.13 | Schematic illustration of a microglomerulus 56
Figure 3.14 | Microglomeruli differ in their presynaptic constituents 57
Figure 3.15 | Microglomeruli differ in their postsynaptic constituents 58
Figure 3.16 | Microglomeruli rearrange during early adult life 61
Figure 3.17 | Overview of automated image analysis of the calyx 62
Figure 3.18 | Microglomeruli rearrange during early adult life 65
Figure 3.21 | Mushroom body phenotypes can be induced using RNAi 71
Figure 3.22 | Automated tracing of LPTC dendrites 74

Abbreviations

2D/3D	two/three dimensional
ACh	Acetylcholine
Apis	*Apis mellifera*
BSA	bovine serum albumin
CA	constitutively active
CNS	central nervous system
cVA	*cis*-vaccenyl acetate
DN	dominant negative
Drosophila	*Drosophila melanogaster*
FCS	fetal calf serum
FL	full length
GABA	gamma-aminobutyric acid
GEF	guanine nucleotide exchange factor
GFP	green fluorescent protein
GTP	guanine triphosphate
LPTCs	Lobula Plate Tangential cells
LTP/LTD	long term potentiation/depression
MARCM	mosaic analysis with a repressible cell marker
mRFP	monomeric red fluorescent protein
NMDA	N-methyl-D-aspartic acid
PBS (PBT)	phosphate buffered saline (+ 0.1 % triton)
PFA	paraformaldehyde
PNS	peripheral nervous system
RFP	red fluorescent protein
RNAi	ribonucleic acid (RNA) interference
STDV	standard deviation
UAS	upstream activating sequence

Acknowledgements

The following work represents my dissertation. The experimental work I describe was partially other people's work and would certainly not have been possible without essential contributions from many colleagues. I did hardly anything entirely on my own and could not have done so. Although I use the pronoun 'I' while describing the experiments it is important to emphasize that I received so much help that this 'I' could be read as 'we' most of the time.

All of my experimental work was carried out at the Max Planck Institute of Neurobiology in Munich in the group of **Gaia Tavosanis** who supervised the study.

Ewa Koper initiated the spine project and provided major contributions. **Jana Lindner** and **Irina Hein** contributed the UAS-mRFP flies and the Dα7 staining presented in Figure 3.6, respectively. Jana Lindner also generated numerous transgenic flies for GAL4/UAS independent labelling of the mushroom bodies (part 3.26). **Malte Kremer** helped with quantifications presented in Figure 3.15, carried out the RNAi pilot screen (part 3.37) and many of the experiments addressing structural plasticity in the mushroom body (parts 3.29, 3.30 and 3.31). **Julia Negele** helped most of all with just being there. I thank all present and previous lab members for technical support and helpful comments.

Frauke Christiansen-Engelhardt and **Stephan Sigrist** helped shaping the ideas for addressing structural plasticity (parts 3.26 and 3.31) and provided transgenic tools (*UAS-Dα7-GFP* and *UAS-D3-cherry* and *mb247-Dα7-GFP*). **Friedrich Foerstner** and **Axel Borst** contributed essential help with the evaluation of the results obtained with automated image analysis (parts 3.27, 3.28 and 3.29). I'm happy that **Claudia Groh, Nancy Butcher** and **Ian Meinertzhagen** collaborated with us to publish the mushroom body calyx anatomy together (part 3.10) and that **Moritz Paehler** and **Peter Kloppenburg** provide essential contributions to our (unpublished) manuscript on structural plasticity.

I enjoyed working with you. Thank you.

Summary

The morphology of dendrites is important for neuronal function and for the proper connectivity within neuronal circuits. The often very complex shape of dendritic trees is brought about by the action of many different genes throughout development. Moreover, neuronal activity is often involved in refining synaptic connections and shaping dendrites. Aiming at a better understanding of the interplay between genes and neuronal activity during dendrite differentiation I was trying to identify suitable neurons in the *Drosophila* central nervous system. Describing the morphology and cytoskeletal organization of a group of visual interneurons involved in motion processing I provided evidence that the dendrites of these neurons do bear small protrusions that share essential characteristics with vertebrate spines. Vertebrate spines received a lot of recent attention because neuronal activity can induce lasting changes in their morphology even in the adult. These morphological changes are believed to be cellular correlates of learning and memory. The observation of similar structures in flies raised the possibility to study structural plasticity in a genetically accessible model organism. Experience-dependent alterations in the volume of a region in the insect brain, called mushroom body calyx, have been shown. The calyx is known to contain the dendrites of olfactory interneurons, Kenyon cells, which are known to be required for the retrieval of olfactory memories in flies. I wanted to address if morphological rearrangements of the dendrites of these cells could underlie the experience-dependent changes in calycal volume. Kenyon cell dendrites and their presynaptic partners are known to form synaptic complexes, called microglomeruli, throughout the calyx. My results help refining the anatomical description of these structures. These findings are important to understand how olfactory experience is represented in the fly brain and how olfactory memories might be formed. Moreover, I developed a computer algorithm to quantitatively describe the morphology of these microglomeruli in an automated manner. Thereby, I found indications for morphological rearrangements of calycal microglomeruli during the first days of the adult life of *Drosophila*. I could show that olfactory experience is not required for these morphological alterations. My findings provide the basis for ongoing attempts to study the influence of neuronal activity on the dendritic morphology of Kenyon cells in more detail.

Introduction

1.1 Dendrites

Neurons are among the morphologically most complex cells and their morphology is closely linked to their function. With their long extended axon and elaborate dendritic arbour, neurons establish the circuitry that detects, stores, and transmits information in the nervous system. Although they come in many shapes and sizes, most neurons have distinct axonal and somatodendritic compartments which are radically different in their signalling properties, cytoskeletal organization and physiological functions. Dendritic trees are usually the morphologically most complex part of neurons and this morphological complexity makes them inherently difficult to study. Therefore, our general understanding of the molecular mechanisms that regulate dendrite growth and branching notably lags behind analogous studies on axon growth and guidance (Baas and Buster, 2004; Dent and Gertler, 2003; Ghysen, 2003; Grueber and Jan, 2004; Horton and Ehlers, 2003; Jan and Jan, 2003; Kim and Chiba, 2004; Libersat, 2005; Libersat and Duch, 2004).

Axonal growth and guidance has been a centre of attention ever since Ramón y Cajal discovered its prominent driving structure, the growth cone. Axons are guided along specific pathways by attractive and repulsive cues in the extracellular environment. Genetic and biochemical studies have led to the identification of highly conserved families of guidance molecules, including netrins, Slits, semaphorins, and ephrins. Guidance cues steer axons by regulating cytoskeletal dynamics in the growth cone (Dickson, 2002; Schnorrer and Dickson, 2004). Similar mechanisms are likely to be important in dendrite growth and guidance as well. Dendrites are involved in collecting information and their morphology reflects this task (London and Hausser, 2005). Depending on the cell type and on the type of input they deal with, dendritic arborization displays a wide range of morphologies, from the single axon-like fibre observed in many sensory neurons to the highly intricate, planar arborization found in Purkinje cells (Cline, 2001; Libersat, 2005; Libersat and Duch, 2004). How are these differences coded for in the developmental program of the neuron? Dendrites progress through several stages of morphogenesis before achieving their mature form. They initiate growth from one or more sites, from either the soma or a proximal segment of the axon. Growing dendrites target a particular receptive territory, within which they branch and achieve a type-specific architecture.

Eventually, branching dynamics slows down and a mature territory and branching complexity is established (Grueber and Jan, 2004; Grueber et al., 2005; Jan and Jan, 2003; Kim and Chiba, 2004; Parrish et al., 2007). Understanding how dendrites accomplish each step of morphogenesis presents an enormously complicated problem. Until recently, the small size of *Drosophila* limited its use in this field. However, methodological advances have overcome some of the challenges of small size (Venken and Bellen, 2005). These advances include the labelling of neuronal populations using the GAL4/UAS system (Brand and Perrimon, 1993) and the visualization and manipulation of individual neurons using the mosaic analysis with a repressible cell marker (MARCM) strategy (Lee and Luo, 1999, 2001). By enabling analyses of identified neurons in undisturbed environments, these advances have opened the complexity of insect dendritic development to combined cellular and genetic analysis (Grueber and Jan, 2004; Grueber et al., 2005; Parrish et al., 2007).

1.2 *Drosophila* **as a model system for dendritic morphogenesis**

Studies on dendrite development in *Drosophila* have provided interesting insights in the recent past. It has been shown that dendritic targeting, branching patterns, territories, and metamorphic remodelling are controlled in specific ways, by intrinsic genetic programs and extrinsic cues, with important implications for function (Grueber and Jan, 2004; Parrish et al., 2007). It was demonstrated that several cell-surface receptors, previously known as axon guidance molecules, are also responsible for the directed outgrowth of dendrites. As known for axons, these molecules play important roles in orienting and positioning of dendrites within the brain and are involved in determining synaptic connectivity as well as the strength of transmission. Such molecules include the ligand-receptor pairs Semaphorin-Neuropilin, Netrin-Frazzled and Slit-Robo (Boyle et al., 2006; Dimitrova et al., 2008; Kim and Chiba, 2004). Many of these studies exploit the experimental advantages of sensory neurons in the *Drosophila* peripheral nervous system (PNS) (Brenman et al., 2001; Dimitrova et al., 2008; Gao and Bogert, 2003; Gao et al., 2000; Grueber and Jan, 2004; Grueber et al., 2005; Grueber et al., 2007; Jinushi-Nakao et al., 2007; Medina et al., 2008; Moore et al., 2002) which have processes that share many essential characteristics with dendrites (Sanchez-Soriano et al., 2005). Besides many striking similarities between the dendrites of these sensory neurons and dendrites in the central nervous system (CNS) there is also a major difference: while synaptic connectivity and dendritic computation pose major constraints on dendritic morphology in the CNS, the dendrites in the PNS lack synaptic input (Sanchez-Soriano et al., 2005). It would be highly desirable to have a model system allowing studies on dendritic morphogenesis and dendritic function in the CNS – ideally with similar genetic accessibility and high resolution imaging as in the periphery. Such a system would allow screening for factors involved in dendrite – and possibly spine (see below) – morphogenesis. It could reveal genetic mechanisms linking synaptic connectivity and dendritic computation with morphology and this possibility represents a major advantage over the PNS dendrites. Ideally, it could ultimately allow studies on the consequences of dendritic manipulations on their function (relating those to behavioural consequences) or on the relation between neuronal activity and dendritic morphology (Chen and Ghosh, 2005; Wong and Ghosh, 2002).

1.3 Candidate neurons for studies on dendrites in the CNS

Aiming to study dendrite morphogenesis in the CNS of the adult *Drosophila* I looked for suitable model systems. The criteria for the selection of candidate neurons were:

a) Complex but stereotyped dendritic morphology,
b) The availability of GAL4 driver lines to allow visualization and manipulation of the target cells specifically,
c) Information on the polarity of the neurons allowing a reliable distinction between dendrites and axons and
d) Information on the function of the neurons.

I will only briefly introduce the pacemaker neurons, the atonal positive neurons and the giant fibre neuron. The neurons I chose to focus on, the Lobula Plate Tangential Cells and mushroom body intrinsic Kenyon cells, are introduced in more detail afterwards.

The pacemaker neurons (LNvs) have complex processes in the optic lobes and can be visualized and manipulated with a very specific GAL4 driver line (Nitabach et al., 2002; Renn et al., 1999). They are part of the circuitry that synchronizes the circadian clocks in the fly (Helfrich-Forster, 2005). Circadian clocks drive rhythmic physiological processes and behaviours in the absence of any rhythmic environmental fluctuations (Hastings et al., 2003; Stanewsky, 2003). In the absence of LNv function (upon induced cell death or electrical silencing) flies do not maintain the rhythmic circadian locomotor activity under constant darkness (Nitabach et al., 2002; Renn et al., 1999). LNv function can thus be tested very easily via behavioural assays – e.g. with available tools to quantify locomotor activity (alternatives are eclosion rythms or larval light avoidance) (Mazzoni et al., 2005). Although their function has been studied extensively it is not entirely clear whether the projections in the optic lobes are dendrites (they might collect input from the rabdomeres to entrain the clock) or axons (they might be required to regulate photoreceptor sensitivity in a circadian manner). It has been demonstrated that LNvs undergo structural rearrangements at a circadian time scale which might allow molecular studies on structural remodelling (Fernandez et al., 2008).

The *atonal (ato)* gene defines a conserved family of genes involved in nervous system development and was initially described as proneural gene that is necessary and sufficient for the development of the *Drosophila* chordotonal organs (Jarman et al., 1993; Wang et al., 2002). An *ato-GAL4* driver was generated and showed specific expression in a cluster of few cells in the optic lobes. Indications for dendritodendritic connection suggested that the ato-expressing neurons are multipolar (Hassan et al., 2000).

The giant fibre circuit is a comparatively simple circuit in the CNS and mediates a well studied behavioural response, the escape jump of the fly (Allen et al., 2006). The giant neuron's dendrites appear to be sufficiently complex and stereotyped to allow detailed morphometric analysis, the polarity is well established and dendritic and axonal processes can be distinguished. Two specific GAL4 driver lines (Allen et al., 1998; Phelan et al., 1996; Trimarchi et al., 1999) are available and allow specific labelling of a single identifiable neuron. The dendrites appear to bear spines and they receive excitatory cholinergic input as shown by acetylcholine (ACh) receptor localization (Fayyazuddin et al., 2006). Due to their large size the neuron is accessible for electrophysiology (Godenschwege et al., 2002).

1.4 Lobula Plate Tangential Cells

Lobula Plate Tangential Cells (LPTCs) are a group of 6 vertical and 3 horizontal cells and their dendrites form a very large and dense dendritic field covering much of the lobula plate (Scott et al., 2002). They provide several major advantages for detailed high resolution imaging: They are large neurons, facilitating visualization, and there exist driver lines allowing cell-specific expression (Joesch et al., 2008; Raghu et al., 2007; Scott et al., 2002). The anatomy of the LPTC dendrites has been described at single-cell resolution and it was demonstrated that each cell is individually identifiable (Scott et al., 2002; Scott et al., 2003a, b). Based on electrophysiological experiments and on the localization of pre- and post-synaptic markers, a dendritic tree, dedicated to receiving input, and an axon can confidently be distinguished (Raghu et al., 2007; Single and Borst, 1998). I considered LPTCs to be the most suitable neurons for detailed genetic analysis of dendrite morphogenesis.

1.1.1 LPTCs are individually identifiable

The morphology of each of the LPTCs has been described in detail (Scott et al., 2002) by MARCM single cell labelling (Lee and Luo, 1999). Position, size and outline of the dendritic field are highly consistent between animals and the morphology of the primary dendrites is stereotyped enough to allow the individual identification of each neuron. It is also possible to identify and trace a single neuron from the entire group of cells and to assemble its morphology from multiple confocal sections (Figure 3.2C).

1.5 Lobula Plate Tangential Cells and the perception of motion

LPTCs are essential to the fly's perception of motion. To guide a fly in three dimensions, flight control crucially relies on optic flow (Frye and Dickinson, 2001). In the fly's brain, the lobula plate encodes visual motion information in a retinotopic manner and is thought to represent a neural control centre for course corrections during flight (Borst and Haag, 2002). LPTCs integrate the output signals of retinotopically arranged local motion detectors (Borst and Egelhaaf, 1992) and connect via descending neurons to the motor neurons in the thoracic ganglia. The electrophysiology of LPTCs has been studied extensively in larger flies (Borst and Haag, 1996; Farrow et al., 2005; Farrow et al., 2003; Haag and Borst, 2002, 2003; Haag et al., 1997; Haag et al., 1999) and recently became accessible in *Drosophila* as well (Joesch et al., 2008).

1.6 Dendritic spines

There have been previous indications of the presence of spine-like processes along the dendrites of LPTCs (Scott et al., 2002; Scott et al., 2003a, b). This raised the interesting question of whether *Drosophila* dendrites might bear spines comparable to the ones described in vertebrates. Dendritic spines have attracted considerable interest because they are suggested to be the cellular effectors of such processes as learning and memory (Yuste and Bonhoeffer, 2001). It is widely assumed that the formation of long-term memories requires activity-dependent long-lasting morphological alterations in plastic neuronal networks, which might take place in the neuronal spines (Alvarez and Sabatini, 2007; Bonhoeffer and Yuste, 2002; Hofer et al., 2006a, b, 2008; Matus, 2005).

This assumption is supported by the findings that spines are motile in hippocampal neurons *in vitro* (Fischer et al., 1998) and that spines are dynamic in acute brain slices, in organotypic cultures and *in vivo* (Dunaevsky et al., 1999; Majewska and Sur, 2003; Nagerl et al., 2004; Oray et al., 2004; Trachtenberg et al., 2002). Dynamic properties of spines are driven by actin (Dunaevsky et al., 1999; Fischer et al., 1998), which is highly enriched in these structures (Fifkova and Delay, 1982; Matus et al., 1982). Regulators of actin dynamics, such as profilin or cofilin are in fact involved in the determination of spine morphology (Schubert and Dotti, 2007). Although the structure of spines is dynamic, several morphological categories have been described around a basic consensus defining them as protrusions of up to 3 μm in length. Such categories include mushroom shaped, branched, thin and stubby spines (Harris et al., 1992).

Due to their morphology, spines represent cellular compartments within which the concentration of Ca^{2+} or the activation of signalling proteins are thought to be independently regulated (Bloodgood and Sabatini, 2007; Hayashi and Majewska, 2005; Yuste et al., 2000). These characteristics would endow spines with the ability to locally specify the effect of single synaptic inputs, a property that could underlie the input-specificity of long-term plasticity.

Indeed, the number of spines can be modified in response to long-term potentiation (LTP) and long-term depression (LTD) induction (Engert and Bonhoeffer, 1999; Maletic-Savatic et al., 1999; Nagerl et al., 2004), *in vitro* paradigms for learning and memory processes (Bliss and Lomo, 1973). More recently, spine number or morphology has been shown to be modified by experience *in vivo* (Holtmaat et al., 2006). Correspondingly, older evidence showed that animals exposed to stimulating environments have an increased number of spines on cortical neurons in comparison to animals grown in non-stimulating environments (Globus et al., 1973). Furthermore, a number of diseases affecting mental function show a clear correlation with the formation of abnormal spines (Fiala et al., 2002b). Thus, spines represent the cellular site at which at least part of the modifications correlated with plastic events happen.

In spite of this intense research on vertebrate spine morphology, dynamics and physiological properties, the molecular analysis of spines (Ethell and Pasquale, 2005; Tada and Sheng, 2006) would benefit from the introduction of a genetically amenable system.

1.7 *Drosophila* and spines

Drosophila has been long established as an important model for the study of learning and memory, allowing the genetic dissection of these processes (Margulies et al., 2005). More recently, genetic analysis of dendrite differentiation in *Drosophila* started providing major contributions to the understanding of dendritogenesis (Parrish et al., 2007). However, a comprehensive analysis of the existence and of the characteristics of dendritic spines in *Drosophila* has never been carried out.

Nonetheless, there are indications that *Drosophila* dendrites might bear spines. In addition to the indications for small dendritic protrusion on LPTC dendrites, spine-like processes were revealed by electron microscopy in lateral horn neurons (Yasuyama et al., 2003).

Taken together, LPTCs appeared to be very promising candidate neurons to study dendrites in the fly CNS because of their advantages for high-resolution imaging and the opportunity to study dendritic spines. Another type of neurons was chosen from the initially identified set of candidates, the intrinsic neurons of the mushroom bodies, the Kenyon cells.

1.8 Kenyon cells

The morphology of Kenyon cell dendrites is less well-defined and more difficult to study than the dendrites of any other candidate neuron mentioned above (Lee and Luo, 1999; Zhu et al., 2003). It is at present impossible to reliably identify the same Kenyon cell in different animals (Murthy et al., 2008). Despite these severe technical disadvantages, I decided to further investigate Kenyon cell anatomy because of the potential to use them for studies on structural plasticity. Structural plasticity is an important aspect of dendritic function and is critical for the establishment of synaptic contacts during development as well as for rearrangements in the adult that are believed to be correlates of learning and memory (Lamprecht and LeDoux, 2004; Lippman and Dunaevsky, 2005; Segal, 2005). While the LPTCs (as well as the giant fibre neurons) might be part of circuits that require little activity-dependent tuning, the Kenyon cells are likely candidates for structural remodelling and thus appeared well suited to complement studies on spines in LPTCs. The neuropil containing the dendrites of Kenyon cells, the mushroom body calyx, was shown to undergo experience-dependent volumetric changes in a number of insect species (Barth and Heisenberg, 1997; Durst et al., 1994; Seid et al., 2005; Withers et al., 1993). Moreover, Kenyon cells are known to be required for the retrieval of olfactory memories and part of the well-characterized olfactory circuit (Fahrbach, 2006; Gerber et al., 2004; Heisenberg, 2003; Keene and Waddell, 2007). These advantages seemed to outbalance the technical challenges. A prerequisite for studies on structural plasticity was a detailed anatomical description of the connectivity in the calyx. I thus investigated the synaptic organization of Kenyon cell dendrites and projection neuron (their presynaptic partners) boutons in synaptic complexes, called microglomeruli, present throughout the mushroom body calyx.

While my neuroanatomical work on the LPTCs was aiming at characterizing morphological properties of individual neurons the studies on the mushroom body calyx focus almost entirely on these microglomerular synaptic complexes. In order to explain the reasons for this choice I will first summarize the anatomy of the olfactory system in flies. This will illustrate that the input to the calycal microglomeruli is comparatively well understood. I will then point out that the opportunities of targeted manipulations in the olfactory circuit represent the main advantage of this system to study structural plasticity. Since experiments addressing structural plasticity appeared less promising in the visual system calycal microglomeruli appeared to nicely complement the analysis of LPTC spines. Moreover, the anatomy of microglomeruli is also of interest to odour representation in the mushroom body and to olfactory memory retrieval.

1.9 Anatomy of the *Drosophila* olfactory system

Current key questions in neurobiology are how sensory information is represented at higher brain centres and how associative memories are established. Central to both questions is the understanding of how the underlying neuronal circuits are organized. In contrast to visual or auditory inputs, a systematic analysis of odour processing has only begun quite recently, revealing a surprising degree of conservation of olfactory circuit design among mammals and insects (Hildebrand and Shepherd, 1997).

Olfaction plays crucial roles in insect survival and reproductive success, mediating responses to food and mates. Insects possess sensitive chemosensory systems that can detect and discriminate among a diverse array of chemicals. The ability to respond to these compounds is conferred by odour and taste receptors, which in both insects and mammals are seven-transmembrane-domain receptors encoded by highly diverse gene families (Hallem and Carlson, 2004; Hallem et al., 2006; Hallem et al., 2004).

Olfactory sensory organs
The fly has two pairs of olfactory organs, the antennae and the maxillary palps. Each antenna contains around 1200 olfactory receptors neurons, whereas each maxillary palp contains about 120 olfactory receptors neurons (Stocker, 1994). Nearly all of the odour receptors expressed in the antenna and maxillary palp have now been characterized, and many of these receptors have been mapped to the functional classes of neurons in which they are expressed (Couto et al., 2005; Fishilevich and Vosshall, 2005). Most, if not all, antennal olfactory receptors neurons express only one functional odour receptor (Hallem et al., 2006; Hallem et al., 2004; Vosshall et al., 1999). Olfactory receptors neurons send axons to the antennal lobe, whose functional organization is remarkably similar to that of the olfactory bulb in vertebrates (Hildebrand and Shepherd, 1997).

Antennal lobes
In the antennal lobe, olfactory receptors neurons synapse onto second order neurons called projection neurons (see Figure 1.1). The antennal lobe can be subdivided into around 50 spherical units called glomeruli. Individual olfactory receptor neurons send axons to only one or a few glomeruli (Stocker, 1994) and individual projection neurons typically innervate only a single glomerulus (Jefferis et al., 2001; Marin et al., 2002; Wong et al., 2002). The glomeruli also contain the processes of local interneurons that branch in multiple glomeruli and that can be either inhibitory (Ng et al., 2002; Stocker, 1994; Stocker et al., 1990; Wilson and Laurent, 2005) or excitatory (Shang et al., 2007). Projection neurons, of which there are around 200, send their axons to both the mushroom body calyx and the lateral horn (Jefferis et al., 2001; Jefferis et al., 2007; Stocker, 1994). Anatomical studies at single cell resolution showed that projection neuron axons have stereotypical branching patterns and terminal areas according to the glomeruli that their dendrites innervate (Jefferis et al., 2005; Jefferis et al., 2001; Jefferis et al., 2002; Jefferis et al., 2007; Jefferis et al., 2004), suggesting that olfactory information might be spatially represented in the higher centres (Komiyama and Luo, 2006) – as also suggested for mice (Zou et al., 2001).

Mushroom bodies
Besides a variety of other functions (Fahrbach, 2006; Gegear et al., 2008; Hong et al., 2008; Joiner et al., 2006; Pitman et al., 2006), *Drosophila* mushroom bodies have been implicated in the generation and retrieval of olfactory associative memories (de Belle and Heisenberg, 1994; Gerber et al., 2004; Heisenberg, 2003; Heisenberg et al., 1985; Keene and Waddell, 2007; Zars

et al., 2000). Approximately 2000 Kenyon cells constitute the intrinsic neurons of the mushroom bodies and receive presynaptic input from projection neurons. Kenyon cells comprise at least (Strausfeld et al., 2003) three subsets based on their axonal projections (α/β, α'/β' or γ) (Crittenden et al., 1998; Lee et al., 1999). Importantly, these anatomical subdivisions correlate with major functional distinctions, such as short-term (γ lobe) and intermediate and long-term (α/β and α'/β' lobes) olfactory memory (Akalal et al., 2006; Krashes et al., 2007; McGuire et al., 2001; McGuire et al., 2003; Pascual and Preat, 2001; Zars et al., 2000). Kenyon cell dendritic projections appear stereotyped within a number of regions in the calyx (Lin et al., 2007; Strausfeld et al., 2003; Tanaka et al., 2004). This suggested that the odour map of the antennal lobe might be retained in a modified form in second-order olfactory centres.

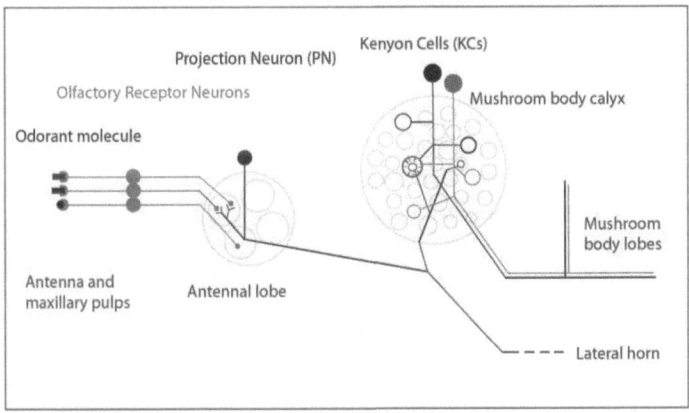

Figure 1.1 | Schematic illustration of the olfactory circuit in *Drosophila*

1.10 Microglomerular complexes in the mushroom body calyx

In spite of considerable recent progress in our understanding of odour representations at the level of the primary olfactory centres, olfactory bulb/antennal lobe, the processing of odour information in higher brain centres remains rather elusive and lacks detailed connectivity information (Keene and Waddell, 2007). However, a detailed understanding of the connectivity between projection neurons and Kenyon cells will be essential to clarify the processing of olfactory input in the mushroom body calyx. Previous electron microscopy studies in *Drosophila* and *Apis* had provided evidence that projection neuron boutons in the adult calyx are surrounded by a number of small post-synaptic profiles including Kenyon cells and few inhibitory interneurons forming large synaptic complexes called "microglomeruli" (Ganeshina and Menzel, 2001; Yasuyama et al., 2002). Labelling with phalloidin demonstrates that in the calyces of crickets, honeybees and ants microglomeruli are enriched in actin (Frambach et al., 2004; Groh et al., 2004). This led to the suggestion that microglomeruli could be sites of structural plasticity. This possibility is supported by their number in honey bees being modified in a temperature-dependent fashion during pupal development (Groh et al., 2006; Groh et al., 2004) and their size increasing in cockroaches as a consequence of olfactory-associative learning (Lent et al., 2007). Mushroom bodies have also been studied in a number of other insects and indications for morphological subdivisions have been described (Fahrbach, 2006; Farris, 2005; Farris et al., 2004; Ganeshina et al., 2006; Gronenberg, 2001; Larsson et al., 2004; Schurmann et al., 2000; Sinakevitch et al., 2001; Sjoholm et al., 2005; Strausfeld, 2002; Strausfeld et al., 1998).

1.11 Microglomeruli and structural plasticity

These earlier findings suggested that calycal microglomeruli *in Drosophila* might undergo experience-dependent alterations. Taken together with the detailed information on the olfactory circuit in flies and its accessibility to genetic manipulation, morphometric analysis of microglomeruli could allow studying the relation between structural rearrangements and the functional alterations they result in. Moreover, the genetic tools available in flies would allow detailed investigations on the genetic basis of structural plasticity. Aiming at ultimately exploiting these advantages, I tried to establish an experimental paradigm to study structural plasticity in the *Drosophila* olfactory circuit.

1.12 Structural plasticity and learning and memory

It is widely accepted that, during learning, reversible physiological changes in synaptic transmission take place in the nervous system, and that these changes must be stabilized or consolidated in order for memory to persist (Dudai, 1996; Keene and Waddell, 2007). The temporary, reversible changes are referred to as short-term memory and the persistent changes as long-term memory. The idea that the creation of stable, persistent long-term memory traces requires gene expression and the resultant synthesis of new proteins is supported by much evidence (Kandel, 2001; Keleman et al., 2007; Krashes et al., 2007). However, molecular changes are transient and so, on their own, are insufficient to explain long-term memory. It is therefore generally believed that structural changes in synaptic morphology, occurring either consequent to protein synthesis or in parallel with it, are also necessary (Lamprecht and LeDoux, 2004; Lippman and Dunaevsky, 2005; Segal, 2005).

Early studies showed alterations in synaptic architecture (such as changes in size or shape) and in the number of synapses after non-associative learning and long-term facilitation in *Aplysia* (Bailey et al., 1992) and in the mammalian hippocampus in response to injury, stimulation or induction of LTP. Similar changes were observed in the neocortex in response to environmental enrichment. Most excitatory synapses in the brain terminate on dendritic spines, which have been the focus of recent work in the mammalian brain (Bonhoeffer and Yuste, 2002; Segal, 2005; Yuste and Bonhoeffer, 2004) (see part 1.6). Modulation of the number of dendritic spines and/or their morphology has been proposed to contribute to alterations in excitatory synaptic transmission during learning (Lippman and Dunaevsky, 2005). Indeed, there is evidence that induction of synaptic plasticity (LTP induction or memory formation) leads to changes in the number or shape of spines (Yuste and Bonhoeffer, 2001).

1.13 Advantages of *Drosophila* for studies on structural plasticity

Although there is much recent progress towards a better understanding of the relation between structural rearrangements and learning and memory, most of the studies remain correlative. The following assumptions nicely fit together: a) alterations in the sensory environment affect neuronal activity, b) neuronal activity can induce structural alterations, and c) structural alterations represent a correlate of learning and memory. However, it remains very challenging to demonstrate the interdependence of these steps in the same circuit. Such an experiment requires a detailed understanding of how the relevant sensory stimuli are represented in the neuronal population undergoing structural modifications. The relation between sensory stimuli and neuronal activity is a focus of current research in the fly olfactory system (Hallem and Carlson, 2004; Hallem et al., 2004; Olsen and Wilson, 2008; Stopfer, 2005; Wilson and Laurent, 2005; Wilson and Mainen, 2006; Wilson et al., 2004). It is believed that olfactory memories are formed just a few synapses away from the sensory organ and it appears conceivable to understand how odours are represented in the mushroom bodies in the near future. Studies on structural plasticity in the calyx could thus ultimately contribute to a better understanding of how morphological rearrangements of a neuron change the information processing it is involved in.

1.14 Olfactory coding in *Drosophila*

The olfactory receptor neurons of the antennae and maxillary palps generate action potentials in response to odour stimulation. The odour responses of many of these olfactory receptor neurons have been characterized through extracellular single-unit recordings from individual olfactory sensilla (de Bruyne et al., 2001). These recordings have revealed that different odorants elicit responses from different subsets of olfactory receptor neurons, and also that olfactory receptor neurons exhibit a remarkable diversity of response properties: responses can be either excitatory or inhibitory and can vary in both intensity and temporal dynamics, depending on the odorant and the olfactory receptor neurons (de Bruyne et al., 2001). Similar olfactory receptor neuron response properties have been described in other insects (Hallem and

Carlson, 2004). Different odorants activate distinct but overlapping subsets of glomeruli in the antennal lobe and the number of activated glomeruli increases with increasing odorant concentration, as revealed by optical imaging (Fiala et al., 2002a; Ng et al., 2002; Wang et al., 2003). Odour coding in the antennal lobe thus appears to involve a spatial map of odorant receptor activation. An electrophysiological analysis of projection neurons similarly revealed that different odorants activate different populations of projection neurons (Broome et al., 2006; Brown et al., 2005; Stopfer et al., 2003; Wilson et al., 2004). In addition, like olfactory receptor neuron responses, projection neuron responses were found to differ in breadth of tuning, signalling mode and response dynamics (Hallem et al., 2006; Olsen and Wilson, 2008; Wilson et al., 2004).

Odour representations in the antennal lobe in flies are thus very similar to odour representations in the mammalian olfactory bulb: Each glomerulus is the site of convergence for axons of olfactory receptor neurons expressing a specific seven-transmembrane-span olfactory receptor. Odorants typically bind multiple olfactory receptors, so the representation of olfactory stimuli is believed to be combinatorial: the activation of distinct groups of glomeruli signifies the presence in the external world of different odorants (Axel, 1995; Buck, 2000; Wilson and Mainen, 2006).

The organization of the circuitry at secondary centres, in both mice and flies, is not well understood. Work in *Drosophila* and locust suggests that, in contrast to the broad odour response tuning of projection neurons, the responses of Kenyon cells to the same odours are usually rare and selective, and electrophysiological studies suggest a model in which Kenyon cells act as coincidence detectors of odour input from projection neurons (Assisi et al., 2007; Ito et al., 2008; Laurent, 2002; Laurent and Naraghi, 1994; Perez-Orive et al., 2002; Stopfer, 2005; Turner et al., 2008; Wilson et al., 2004). Individual odours could be represented as sparse labelled lines in the mushroom bodies and this belief is central to current models of odour memory (Heisenberg, 2003; Keene and Waddell, 2007). However, it is worth noting that although odours may evoke activity in a sparse array of mushroom body cell bodies, and perhaps dendrites, it is not clear how the information is represented in the mushroom body lobes because the extent of interconnection by gap junctions and/or chemical synapses is unknown (Keene and Waddell 2007).

Despite these severe gaps in our understanding of olfactory perception in the fly, the relative simplicity of the circuit and the accessibility to experimental manipulation represent major advantages over other sensory systems and other organisms.

1.15 Indications for structural plasticity in insects

It has been demonstrated that the volume of the mushroom body calyx is sensitive to rearing conditions and visual experience in flies, bees and ants (Durst et al., 1994; Farris and Strausfeld, 2003; Groh et al., 2006; Groh et al., 2004; Ismail et al., 2006; Kuhn-Buhlmann and Wehner, 2006; Seid et al., 2005; Withers et al., 1993). This suggests that at least one of the neuronal (or glial) populations in the mushroom body rearrange upon sensory manipulations. Kenyon cells are the intrinsic neurons of the mushroom bodies and it has been observed that spine density in Kenyon cells is affected by sensory experience in *Apis* (Farris et al., 2001). Mutants impaired in olfactory associative learning have been shown to be defective in experience-dependent volume expansion of the calyx in *Drosophila* (Barth and Heisenberg, 1997; Hitier et al., 1998). Calycal morphology is modified upon olfactory associative learning in cockroaches (Lent et al., 2007). Taken together, these observations encouraged me to investigate if Kenyon cells in *Drosophila* undergo experience-dependent structural rearrangements.

1.16 Genetic analysis of dendrite and spine morphogenesis

The main reason to use *Drosophila* to characterize dendrites and spines and to aim at establishing a model system allowing investigations on structural plasticity is to ultimately explore the unique opportunities for genetic analysis in the fly. The *Drosophila* PNS has been extremely useful for the genetic analysis of dendrite morphogenesis (Grueber and Jan, 2004; Grueber et al., 2005; Jan and Jan, 2003; Parrish et al., 2007). However, the sensory neurons in the periphery do not receive synaptic input and they thus lack an essential property of dendrites. I asked if the anatomical information obtained for LPTCs and Kenyon cells could be used for genetic analysis of dendrite and spine morphogenesis and allow circumventing the constraints of PNS dendrites. I aimed at establishing experimental conditions allowing genetic manipulations of LPTCs and Kenyon cells using UAS constructs for RNAi mediated knockdown of candidate genes and developed computer based tools for automated morphometric analysis.

1.17 Genetic screens

Geneticists have traditionally sought to gain insight into complex biological processes through forward genetic screens (Jorgensen and Mango, 2002; Nusslein-Volhard and Wieschaus, 1980; Page and Grossniklaus, 2002; St Johnston, 2002). Mutations are generated at random, phenotypes of interest are scored, and the mutated gene is subsequently identified. This approach has been remarkably successful, particularly in *Drosophila*. Traditional forward screens, such as the Nobel-prize-winning screen for embryonic-patterning mutants (Nusslein-Volhard and Wieschaus, 1980), require labour-intensive mapping of randomly generated mutations. To circumvent these limitations a number of alternative means of mutagenesis have been developed. These include gene disruption with transposable element (Spradling et al., 1999) or chromosomes containing large deletions or gain of function approaches such as enhancer-promoter (EP)-induced overexpression of random loci. Alternatively, the GAL4/UAS system (Brand and Perrimon, 1993) can be used to restrict genetic manipulations to a particular cellular population and to a particular time in development in *Drosophila*. It is thus possible to circumvent lethality resulting from an essential function of the target gene at an earlier developmental time point or in another tissue. A large number of UAS constructs allowing cell specific manipulation of target genes is available. These include dominant-negative, constitutively active or overexpression UAS constructs. Moreover, UAS constructs for RNAi mediated gene knockdown have been generated for >80% of the genome (Dietzl et al., 2007).

1.18 Genetic analysis of dendrite and spine morphogenesis

These tools are critical for genetic analysis of dendrite and spine morphogenesis because they allow genetic manipulations of target genes in selected neurons developing in their endogenous context: the intact brain. Since the morphological characterization of dendrites requires high-resolution imaging and sophisticated morphometric analysis the efficacy and specificity of the genetic manipulation are important – especially if large sets of candidate genes should be tested.

I briefly summarized my reasons to consider studies on dendrites important and *Drosophila* a good model organism in this context. The candidate neurons I choose to focus on, Lobula Plate Tangential Cells and Kenyon cells, were introduced with a focus on the aspects of dendritic function they may help to illuminate. LPTCs appear well suited for studies on dendritic morphogenesis because they allow high-resolution imaging of uniquely identifiable dendrites. Since LPTCs bear structures that closely resemble vertebrate spines these small dendritic protrusions received special attention. Spines are considered to be targets of structural remodelling processes associated with learning and memory. Kenyon cells were introduced with an emphasis on the olfactory system they are part of because the detailed knowledge about this circuit and the genetic tools to manipulate it constitute main advantages for studies on structural plasticity. Finally, the advantages of genetic screens were outlined because the prospects of a genetic screen on dendritogenesis in the *Drosophila* CNS were tested.

Material and Methods

2.1 Buffers

Name	Ingredients
PBS (10x)	100mM Na_2HPO_4; pH 7.4
	20mM KH_2PO_4
	1.37 M NaCl
	27mM KCl
PBT	0.05% Triton X-100 in 1x PBS
Blocking buffer for antibody staining	10% fetal calf serum in PBT
PFA (4%)	4% Paraformaldehyde in 1x PBT
Fly food (1 L)	yeast 15 g
	agar 11.7 g
	molasses 80 g
	corn flour 60 g
	methylparaben 2.4 g
	propionic acid 6.3 ml
Fly water	0.8% CH_3COOH in dd H_2O

2.2 Fly stocks

The following list of fly stocks only summarizes frequently used and important stocks; composite stocks of the elements listed below are not included. In addition, a number of stocks were used to manipulate candidate genes in the LPTCs, in the mushroom bodies or both. These include mutants and UAS constructs (overexpression, dominant negative and constitutively active variants) as well as UAS RNAi lines. Most of those lines were obtained from the Bloomington stock centre (Bloomington, USA) or the VDRC stock centre for RNAi lines (Vienna, Austria) and are not listed below (see parts 3.35 and 3.36). Only lines of potential interest for future studies that were obtained from other labs are included.

Name	Source
db331-GAL4	Alexander Borst (Germany)
3A-GAL4	Alexander Borst (Germany)
ok107-GAL4	Bloomington Stock Centre (USA)
c739-GAL4	Bloomington Stock Centre (USA)
1471-GAL4	Bloomington Stock Centre (USA)
17d-GAL4	Scott Waddell (USA)
201y-GAL4	Bloomington Stock Centre (USA)
c305a-GAL4	Scott Waddell (USA)
gh146-GAL4	Thomas Hummel (Germany)
mz19-GAL4	Thomas Hummel (Germany)
np3529-GAL4	Thomas Hummel (Germany)
gad1-GAL4	Ron Davis (USA)
repo-GAL4	Bloomington Stock Centre (USA)
ato-GAL4	Bloomington Stock Centre (USA)
pdf-GAL4	Bloomington Stock Centre (USA)
ok307-GAL4	Bloomington Stock Centre (USA)
mb247-Dalpha7-GFP	Stephan Sigrist (Germany)
mb247-abp-KO	Julia Negele, Jana Lindner (Tavosanis lab)
mb247-actin-GFP	Jana Lindner (Tavosanis lab)
mb247-actin-KO	Jana Lindner (Tavosanis lab)
mb247-dsRed	André Fiala (Germany)
UAS-actin-GFP	Hiroki Oda (Japan)
UAS-abp-KO	Julia Negele, Jana Lindner (Tavosanis lab)
UAS-tubulin-GFP	Nicole Grieder (Switzerland)
UAS-mCD8-GFP	Bloomington Stock Center (USA)
UAS-RFP	Jana Lindner (Tavosanis lab)
UAS-myrmRFP	Bloomington Stock Center (USA)
UAS-synaptobrevin-GFP	Bloomington Stock Center (USA)
UAS-mCD8-cherry	Takashi Suzuki (Germany)
UAS-D3-strawberry	Stephan Sigrist (Germany)
UAS-Dalpha7-GFP	Stephan Sigrist (Germany)
UAS-dicer2	Barry Dickson (Austria)
UAS-GMA	Daniel Kienhart (USA)
UAS-CaMKII.T287A	Leslie Griffith (USA)
UAS-CaMKII.T287AD	Leslie Griffith (USA)
UAS-CaMKII-eYFP	Sam Kunes (USA)
UAS-CaMKII-eYFP 3'UTR	Sam Kunes (USA)
UAS-Nut3-eYFP	Sam Kunes (USA)
UAS-Rdl-HA	Alexander Borst (Germany)
UAS-sra.RNAi	Christian Klämbt (Germany)
UAS-mef2(EP)	Justin Blau (USA)
UAS-appl.RNAi	Bassem Hassan (Belgium)

2.3 Antibodies and probes

Name	Source
nc82 (mouse; 1:20)	Thomas Hummel (Germany)
Dalpha7 (rat; 1:2000)	Hugo Bellen (USA)
mCD8 (rat; 1:100)	Caltag Laboratories (USA)
GFP (rabbit; 1:1000)	Invitrogen (Germany)
vGAT (rabbit; 1:1000)	Dick Nässel (Sweden)
ChAT (mouse; 1:100)	DSHB (USA)
Synaptotagmin (rabbit; 1:25)	Hugo Bellen (USA)
anti-mouse Rhodamin X (goat; 1:100)	Jackson Laboratories (USA)
anti-rabbit Rhodamin X (goat; 1:100)	Jackson Laboratories (USA)
anti-mouse Alexa 488 (goat; 1:100)	Invitrogen (Germany)
anti-mouse Cy3 (donkey; 1:100)	Jackson Laboratories (USA)
anti-GFP (rabbit; 1:1000)	Living colors (USA)
phalloidin Alexa 568	Invitrogen (Germany)

2.4 Fly genetics

Flies were raised at 25°C with 70% relative humidity on standard cornmeal agar medium. The *db331-GAL4*, *UAS-actin-GFP* (Verkhusha et al., 1999), *UAS-GMA* (Edwards et al., 1997) *UAS-tubulin-GFP* (Grieder et al., 2000) and *UAS-Dicer-2* (Dietzl et al., 2007) lines were obtained from A. Borst, H. Oda, D. Kiehart, N. Grieder and B. Dickson, respectively. The *17d-GAL4*, *c305a-GAL4* and *UAS-eag* lines were kindly provided by S. Waddell (University of Massachusetts Medical School, Worcester, MA, USA), *UAS-actin-GFP* by H. Oda (JT Biohistory Research Hall, Osaka, Japan), *gh146-GAL4*, *np3529-GAL4* and *mz19-GAL4* by T. Hummel (Universität Münster, Münster, Germany) and *gad1-GAL4* by R. Davis (Baylor College of Medicine, Houston, TX, USA). The *UAS-mCD8-GFP* (Lee and Luo, 1999), *UAS-myr-mRFP*, *UAS-rac1.N17*, *UAS-rac1.L* (Luo et al., 1994) *1471-GAL4*, *repo-GAL4*, *UAS-mCD8-GFP*, *ok107-GAL4*, *c739-GAL4*, *201y-GAL4*, *UAS-myr-mRFP*, *UAS-shi* and *UAS-dORK* lines were obtained from the Bloomington Stock centre.

For RNAi knockdown of *rac1*, either line #49247 obtained from the VDRC or UAS-mRed as a control were crossed to *db331-UAS-mCD8-GFP; UAS-dicer2*. The progeny was raised at 27°C throughout development. MARCM experiments were performed as described previously (Lee and Luo, 1999) and using the following genotype: *hsFLP, elavGAL4, UAS-mCD8-GFP/ +; FRT42D, tubGAL80/FRT42D*. Late third instar larvae were heat-shocked for 40 min at 38°C.

To allow GAL4/UAS independent labelling of the mushroom bodies, transgenic flies carrying several genes under the control of the mushroom body specific *mb247* enhancer (obtained from

Andreas Thum, Fribourg, Switzerland) were generated. *mb247-actin-GFP* and *mb247-actin-KO* and *mb247-abp-KO* as well as *UAS-actin-KO* and *UAS-abp-KO* were generated together with Jana Lindner and Julia Negele. The *mb247-Dalpha7-GFP* line was generated by Frauke Christiansen-Engelhardt in the Stephan Sigrist laboratory.

For genetic analysis of dendrite and spine morphogenesis a stock containing *db331-GAL4* and *ok107-GAL4* and *UAS-actin-GFP* (Verkhusha et al., 1999) as a reporter was generated. The actin-GFP reporter was chosen because actin specifically localizes to spines in the LPTCs and to claw-like endings in Kenyon Cells and thus highlights small dendritic structures facilitating screening for subtle morphological alterations. All three constructs in the stock were homozygous and female progeny from virgins crossed to males carrying UAS constructs to manipulate candidate genes could thus be screened (carrying all four constructs in heterozygous condition regardless of the insertion site of the UAS construct). Since only males were required from the UAS lines the genetic preparations were trivial and could easily be handled on a large scale. Since the efficacy of RNAi downregulation was reported to be increased upon overexpression of Dicer2 an adequate variant of the stock (*db331-GAL4; UAS-actin-GFP; UAS-dicer2; ok107-GAL4*) was assembled. To allow investigating either the LPTCs of the Kenyon cells alone the following stocks were prepared: *db331-GAL4 UAS-mCD8GFP; UAS-dicer2* and *db331-GAL4; UAS-actin-GFP; UAS-dicer2* and *db331-GAL4 UAS-GMA; UAS-dicer2* and *UAS-actin-GFP; ok107-GAL4* and *UAS-actin-GFP; UAS-dicer2; ok107-GAL4* and *mb247-Dα7-GFP*.

2.5 Immunohistochemistry and confocal imaging

Brains were dissected in phosphate buffered saline (PBS), fixed for 40 minutes (larval brains for 15 minutes) in 4% formaldehyde in PBS and rinsed in PBS (Wu and Luo, 2006). They were then whole-mounted in VectaShield (Vector Laboratories, Burlingame, USA) on a slide and covered by a coverslip using double-sided tape as spacer. Brains were mostly oriented to lay with their antennal lobes down and calyces up to obtain the plane of imaging illustrated in Figure 3.9A. For immunohistochemistry brains were blocked in 10% fetal calf serum (FCS) in PBT (0.1% Triton X-100 in PBS) for 30 minutes unless otherwise stated. All confocal fluorescence microscopy was done with a Leica TCS SP2 confocal microscope (Leica Microsystems, Wetzlar, Germany) using a 63x/1.4 oil-immersion objective.

Brains from *db331-GAL4/+; UASmCD8/+* or w *db331-GAL4; UAS-actin-GFP/+* adult females (4-8 days old) were stained as reported previously (Wong et al., 2002) using the following primary antibodies: mouse anti-Bruchpilot (nc82, 1:20, kindly provided by T. Hummel), rabbit anti-synaptotagmin (1:25, kindly provided by H. Bellen). To detect the endogenous localization of the Dα7 acetylcholine receptor subunit brains were fixed in 4% formaldehyde for 5 minutes at room temperature (S. Raghu, personal communication). Primary antibodies were anti-Dα7 (a kind gift of H. Bellen, 1:2000) and anti-GFP (Living colours, 1:1000) diluted in blocking solution (10% FCS and 0.3% Triton X-100 in PBS).

For immunohistochemistry in the mushroom bodies I used the following probes: 1:25 Alexa Fluor 568 phalloidin to visualize filamentous actin (Invitrogen, Karlsruhe, Germany) for 4 h at room temperature, 1:25 α-synaptotagmin raised in rabbit (Littleton et al., 1993) (pre-absorbed with *Drosophila* embryos, kindly provided by H. Bellen, Howard Hughes Medical Institute, Houston, USA) or 1:100 ChAT4B1 (DSHB, Iowa, IA, USA) overnight at 4°C. For nc82 labelling (Wagh et al., 2006) (1:50; kindly provided by T. Hummel, Universität Münster, Germany) I used PBT with 0.3% Triton throughout the procedure and incubated for 2 d. After washing with PBT the following secondary antibodies were used: 1:100 goat α-rabbit conjugated with Rhodamine Red X or 1:100 donkey α-mouse conjugated with Cy3 (both Jackson Laboratories, Suffolk, England). I did not obtain similar labelling with secondary antibodies alone.

For genetic analysis of dendrite and spine morphogenesis a visual screening procedure was developed. The goal was to identify potentially interesting genes without any detailed investigation. Since this step had to be done routinely it was optimized for speed and efficacy at the expense of possibly significant numbers of false negatives and false positives. As I aimed at the rapid identification of candidate genes rather than a conclusive characterization of any set of genes I considered false negatives acceptable and unavoidable. False positives could be identified in more careful secondary investigations.

5-10 female progeny per genotype were dissected and fixed 2-4 days after eclosion and mounted onto coverslips (Wu and Luo, 2006). No immunolabelling was required. Visual investigation at the confocal microscope to assess general morphology of the LPTCs and mushroom bodies was done within a couple of seconds per brain. High magnification images of the medial VS1 region and the calyx were only obtained when needed and compared to control images. Spine densities of selected promising candidates (*rac1*, *sra1*, *fmr1*) were quantitatively analyzed as described below. No quantitative measurements on calycal phenotypes were done but could easily be obtained with the recently developed automated image analysis tools.

2.6 Manual quantitative morphological analysis

Images were processed using Adobe Photoshop CS2 and illustrations were assembled using Adobe Illustrator CS2 (Adobe, San Jose, USA). 3D reconstructions were generated with Amira (Visage Imaging, Berlin, Germany). Quantitative measurements were done with Amira, ImageJ (http://rsb.info.nih.gov/ij/index.html), ScionImage (Scion Corporation, Frederick, USA) or

Definiens (Definiens, Munich, Germany) as indicated. Statistical analysis was done with Excel (Microsoft, Redmont, USA) or MatLab (MathWorks, Natick, USA).

Quantification of LPTC spine density and length was done for the following genotypes: *db331-GAL4 UAS-mCD8-GFP/+; UAS-myr-mRFP/+, db331-GAL4/+; UAS-actin-GFP/+; UAS-myr-mRFP/+, db331-GAL4/+; UAS-GMA/+; UAS-myr-mRFP/+, db331-GAL4 UAS-GMA/+; UAS-myr-mRFP or UAS-rac1.L or UAS-rac1.N17/+*. Image stacks of small dendritic branchlets of VS1 were taken with a Leica SP2 confocal microscope. Branchlet length and spine length were measured with ImageJ on projections of confocal stacks. Spine numbers were counted on 3D reconstructions generated in Amira. All protrusions between 0.2 and 3 µm length were considered as spines. Spine density and length were calculated for individual flies. Five animals were analyzed and averaged per data point. Images of 5-10 dendritic branchlets were quantified for each animal. Data for the branchlets or individual spines was averaged to obtain the value per animal for spine density and length. Analysis of spine morphology classes was also done on the middle region of VS1 neurons of *db331-GAL4/+; UAS-GMA/+; UAS-myr-mRFP/+* or *db331-GAL4 UAS-mCD8-GFP/+* female flies. Projections from z stacks of confocal images were processed in ImageJ. First, the total number of spines present on a restricted fragment of VS1 was counted and labelled. Then, spines were assigned to one out of four classes following the criteria described in the text. Spine category distribution was determined based on more than 100 spines per animal and then averaged for the five animals analyzed. The presented data were processes by volume rendering (using Amira, Figure 3.4B).

To quantify actin or tubulin enrichment in dendritic spines, the brightness level of 15-200 sample areas of same size in dendrites or in spines was measured in 4 or 5 representative images using ImageJ.

To estimate the percentage of spines that contain ectopically expressed ACh-receptor the following genotype was analyzed: *db331-GAL4/+; UAS-Dα7-GFP/+; UAS-myr-mRFP/+*. Spines were identified morphologically using the myr-mRFP signal and were assigned as ACh receptor positive or negative based on the Dα7-GFP signal. >280 spines from 5 animals were classified; percentages were determined per animal and then averaged.

Spine density upon ACh receptor overexpression (*db331-GAL4/+; UAS-Dα7-GFP/+; UAS-myr-mRFP/+*) was compared to *db331-GAL4/+; UAS-GMA/+; UAS-myr-mRFP/+*. The myr-mRFP signal was used in both cases to count the spine number. Due to limitations in image quality projections through confocal stacks were used instead of 3D reconstructions as described above. >260 spines from five animals of each genotype were analyzed. The statistical analysis was done as described above.

Juxtaposition (<0.1 µm) of a presynaptic (nc82) staining to spines or ectopically expressed ACh receptor (*db331-GAL4/+; UAS-Dα7-GFP/+*) was evaluated using 3D reconstructions generated with Amira from confocal sections. Spines or receptor patches were classified as juxtaposed or non-juxtaposed to presynaptic staining upon rotation of the 3D reconstruction.

Random distribution of presynaptic labelling was estimated by rotating the nc82 channel by 90° relative to the Dα7-GFP channel. >1300 receptor patches and >250 spines, respectively, from 5 animals were analyzed. The statistical analysis was done as described above.

To analyze the anatomy of the mushroom bodies I acquired confocal z stacks in 0.12 μm slices and used Amira software to generate 3D models, sagittal and coronal planes. The single Kenyon cell clones in Figure 3.9C and Figure 3.10C are 3D representations projected onto a relevant single confocal section showing phalloidin or synaptotagmin labelling, respectively.

Synapse distribution was quantified on 225 μm² large medial fractions of single confocal sections taken at a medial calycal section (Figure 3.8). Synapses within the centres of microglomeruli were identified by the lack of overlap between presynaptic (nc82) labelling and postsynaptically expressed actin-GFP (very similar labelling were obtained with the anti-synaptotagmin antibody). A total of 2950 presynaptic puncta was analyzed. Random distribution of presynaptic labelling was estimated by rotating the nc82 or the synaptotagmin channel by 90° relative to the actin-GFP channel (Lang et al., 2007). 12 optical sections were quantified in each case.

3D models were used to estimate the number of GABAergic cell bodies in the proximity of the calyx and to reveal that glial processes do not enwrap subcompartments of the calyx. They appear to form a meshwork-like structure instead.

Confocal z stacks were imported into ImageJ to measure and count microglomeruli. By constantly moving through the z stack, I tried to identify the plane of maximal extension (in xy) of each microglomerulus and marked it to avoid double counting. Complete labelling of microglomeruli was technically difficult to achieve; I estimate from several counts that 20% of all microglomeruli were systematically missed in the routine counts or too small to be easily identifiable. I counted 786 ± 73 microglomeruli (n=3 calyces, error is standard deviation, STDV) from 24 hours old animals, therefore estimating the total number to be around 1000. Similarly, I counted the microglomeruli of three calyces of 14 d old flies (577 ±19) and estimate the total number at that age to be around 750.

To quantify the percentage of microglomeruli that contain *gad1*-positive processes, γ-, α'β'- or αβ-neurons I used genetic labelling with mCD8GFP (driven by the *gad1-GAL4* driver) or actin-GFP (driven by the *1471-GAL4*, *c305a-GAL4* or *c739-GAL4* drivers, respectively). I co-labelled with phalloidin and counted actin-GFP positive versus actin-GFP negative microglomeruli (Figure 3.15) as described earlier. Since the visualization of the total population thus relied on phalloidin labelling instead of genetic labelling image quality was not as high as in the earlier counts. Moreover, the analysis was complicated by the necessity to assign microglomeruli into actin-GFP positive or negative categories while using GAL4 drivers that differed in expression intensity. I thus looked at >5 brains to estimate the percentages and to ensure that the examples were representative and then counted >500 microglomeruli from only one brain per genotype to confirm the estimate. For each identified microglomerulus I allocated a position in the xy (optical) plane using ImageJ and plotted this positional information while neglecting the different positions in z to obtain the illustrations in Figure 3.15D-F.

For the quantification of the inner microglomerular surface, medial confocal sections of the mushroom body calyx were obtained. The 25 biggest and best-defined microglomeruli were visually identified and their inner surfaces were traced using Adobe Photoshop. Images were converted into binary images and imported into ScionImage to determine the surface of the traced entities. Average values of five animals were used to obtain the specified data points. The following genotypes were analyzed: *UAS-actin-GFP/+; myr-mRFP/ +; ok107-GAL4/+ or actin-GFP/+; UAS-dORK1ΔC1/+; ok107-GAL4/ + or actin-GFP/eagΔ932; ok107-GAL4/ +, UAS-actin-GFP/ +; ok107-GAL4/ +*. All quantitative analysis was done blind.

2.7   Automated quantitative morphological analysis

mb247-Da7-GFP females and males were imaged for all experiments quantified with automated image analysis. Costaining with nc82 was used to label presynaptic sites. Confocal stacks of 50 images (512x512 pixels) and spanning the calyx in steps of 0.3-0.6 um were taken (Leica SP2; 63x objective) and used to determine calycal volume and microglomerular morphology. Neuropil volume was determined from confocal stacks spanning the brain in 50 steps of 1.5-2.5 um (512x512 pixels; 20x objective).

Definiens software was used for fully automated image analysis of microglomerular complexes. A full description of the algorithm as well as comments on the evaluation of the algorithm are omitted because they will appear in a primary research article that is currently in preparation.

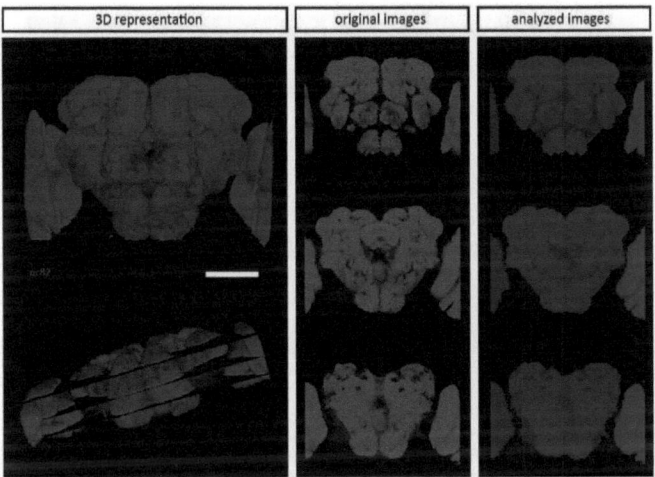

Figure 2.1 | **Overview of automated image analysis of brain volume**

A 3D reconstruction (Amira) of the brain (left). Confocal stacks of 50 sections spanning the brain were used for morphometric analysis. 3 such sections are illustrated and representative original images (nc82 immunolabelling, red) are shown in (middle panels). All images were analyzed fully automatically using Definiens software (right panels). Neuropil volume was calculated as the sum of pixels assigned as neuropil (red) from all sections. Optic lobes (blue) were excluded because their volume was reported to depend on rearing conditions. Their outlines were identified based on their proximity to the image borders and the gap in nc82 staining between optic lobes and central brain. Scale bar: 100 μm.

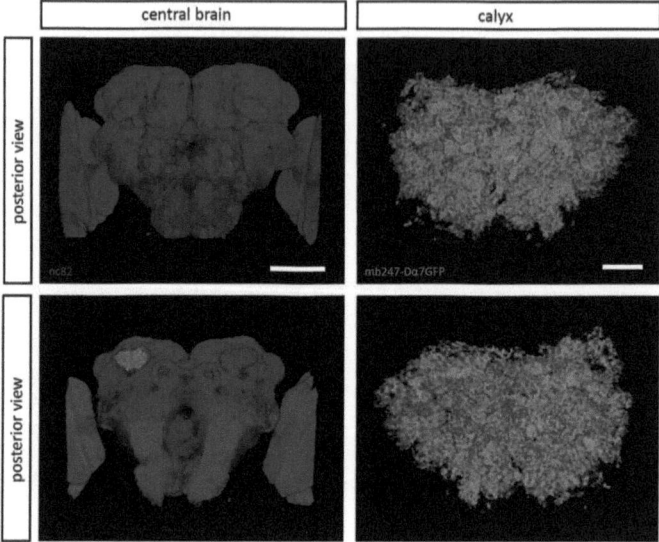

Figure 2.2 | **Illustration of relative volumes of calyx and brain**

3D reconstructions (Amira) of the brain (left panels, red) and calyx (right panels, green) illustrating their relative proportions. The volumes of the central brain (excluding the optic lobes) and of the calyx were determined. Calycal volumes were used for normalization purposes because of a higher consistency of the measurements throughout the data set. Scale bars: 100 μm (left) and 10 μm (right).

Results

3.1 Model systems for dendrites in the fly CNS

Aiming at studying dendrite morphogenesis in the CNS of the adult *Drosophila* I looked for suitable model systems. Specific GAL4 drivers (*pdf-GAL4*, *ato-GAL4*, *db331-GAL4* and *ok107-GAL4*, respectively) were used to express mCD8-GFP in candidate neurons to investigate their morphology (Figure 3.1). The GAL4 driver lines for pacemaker neurons (*pdf-GAL4*) (Renn et al., 1999) and LPTCs (*db331-GAL4*) (Raghu et al., 2007) appeared very specific and allowed the visualization of very few neurons with good expression levels and low background. High resolution imaging of atonal-positive neurons (*ato-GAL4*) (Hassan et al., 2000) was also possible but more challenging. *ok107-GAL4* (Yang et al., 1995) was strongly and specifically expressed in the vast majority of Kenyon cells but none of the available GAL4 lines allowed the visualization of single Kenyon cells. While published information on LPTCs and Kenyon cells allowed a reliable distinction between axons and dendrites the polarity of the processes in atonal-positive neurons and pacemaker cells was less well-defined. The presynaptic marker Synaptobrevin-GFP was expressed in both cell types. Synaptobrevin-GFP was not clearly restricted to distinct processes suggesting axonal or bipolar character of the processes of pacemaker cells and atonal-positive neurons. I therefore decided to restrict my analysis to the LPTCs and Kenyon cells.

The LPTCs appeared to be promising candidates to study dendritogenesis genetically and the characterization of their anatomy and cytoskeletal organization will be presented in the next paragraphs. Small processes along their dendrites were found to share characteristics with vertebrate spines and a systematic evaluation of these structures will be the focus of the anatomical work on the LPTCs. Kenyon cells were chosen because of their implication in learning and memory and the better prospects to study structural rearrangements; their anatomy and the attempts to study structural plasticity will be presented subsequently. In retrospect both systems seem to be fortunate choices. A recent publication (Fayyazuddin et al., 2006) and own observations (the dendrites bear spine-like protrusions that are actin-enriched; Ewa Koper and F.L.) suggest that large cells of the giant fibre system could be an interesting supplement to the LPTCs.

3.2 Lobula Plate Tangential Cells

The dendrites of the LPTCs form a very large and dense dendritic field covering much of the lobula plate. Position, size and outline of the dendritic field are highly consistent between animals and the morphology of the primary dendrites is stereotyped enough to allow the individual identification of each neuron. It is also possible to identify and trace a single neuron from the entire group of cells and to assemble its morphology from multiple confocal sections. This allows assessing single cell morphology as previously described by MARCM single cell labelling (Scott et al., 2002; Lee and Luo, 1999) (Figure 3.2C).

The most distal of the vertical neurons, VS1 is easiest to identify and there is minimal overlap with other LPTC dendrites at a medial position of the dendritic tree. I thus decided to focus on the medial part of VS1 for all quantitative analysis of fine dendritic branches and spines to minimize variation. Secondary branches from the VS1 primary dendrite only branch off distally and often bifurcate. Their exact position and length shows some variation but the finest branches always give rise to an oval outline of the entire dendritic field. The small branches emerging from the VS1 primary dendrite are covered with small spine-like protrusions that I will call spines for simplicity from now on. Spine density and morphology within the selected region of VS1 are roughly consistent between animals. The high degree of stereotypy and the possibility to restrict the analysis to a single identifiable neuron is a major strength of the system and a prerequisite for molecular studies on morphogenesis of dendrites and spines.

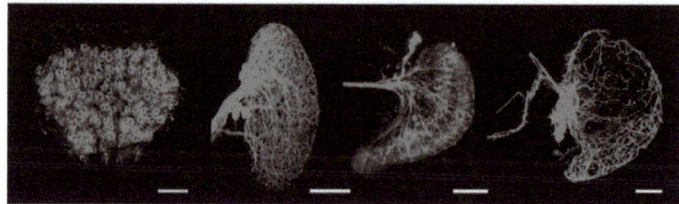

Figure 3.1 | **Different dendrite model systems**

Representative images of the mushroom body calyx, the Lobula Plate Tangential Cells, the *atonal*-positive neurons and the pacemaker cells (from left to right). The *ok107-GAL4*, *db331-GAL4*, *ato-GAL4* and *pdf-GAL4* drivers were used, respectively, to drive actin-GFP (left image only) or mCD8-GFP. Scale bars: 10 µm (left) or 50 µm.

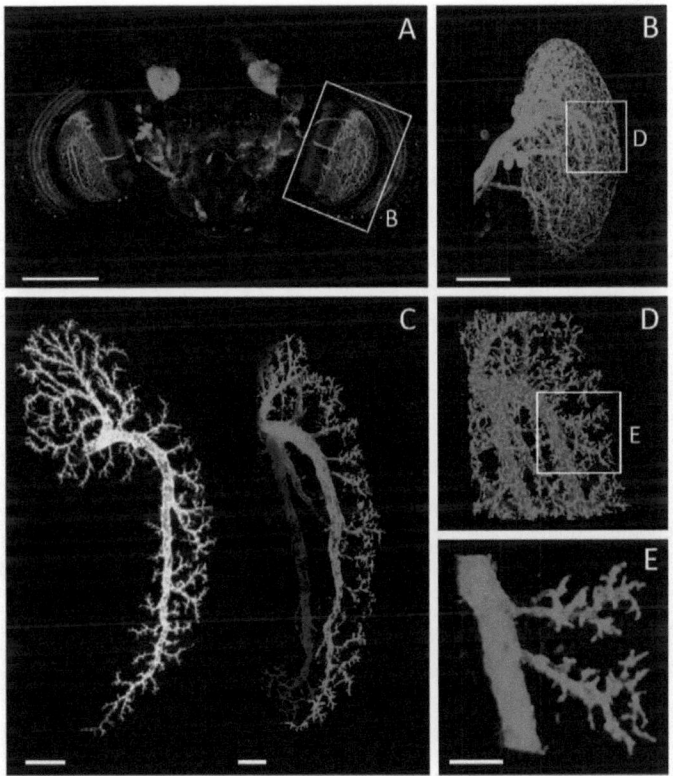

Figure 3.2 | **LPTC overview**
(A) Single confocal section through the entire brain illustrating the localization of LPTCs in the optic lobes (boxed region, magnified in B). (B) Projection through a confocal stack spanning the LPTC dendritic field. (C, left) A single LPTC (VS1) visualized with MARCM (taken from (Scott et al., 2002) or assembled from various confocal sections (C, right). (D, E) 3D reconstructions of the boxed regions in B and D, respectively. Scale bars: 150 µm (A), 50 µm (B), 10 µm (C) and 5 µm (E).

To analyze in detail the morphology of LPTC dendrites animals expressing mCD8-GFP (Lee and Luo, 1999) (Figure 3.3A-E) and cytoplasmic mRFP (monomeric Red Fluorescent Protein; (Campbell et al., 2002) (Figure 3.3A, C, E) under the control of *db331-GAL4* were generated. Spines could be detected on the dendrites of both horizontal and vertical cells (Figure 3.3A, B) similarly to what was reported by Scott et al. (Scott et al., 2002; Scott et al., 2003a, b). These spines were present on all dendrite orders, with 5-10% being located on the primary dendritic branches, but their density was highest on fine branchlets (third order dendrites and higher; Figure 3.3C, D, arrows).

3.3 LPTC dendrites bear spine-like processes

The length and density of spines was quantified in projections of confocal stacks and an average length of the processes of 1.1 µm (± 0.1 µm) was obtained. Spine numbers were counted on volume-rendered 3D reconstructions of confocal microscope optical section series of dendritic branches, which allowed observing even short processes with high resolution. An average density of 1.2 spines/µm (± 0.1 spines/µm) was calculated, which is about twice as high as previously reported for single cell clones (Scott et al., 2002; Scott et al., 2003b). This is most likely due to the use of 3D reconstructions, since similar values were obtained as in (Scott et al., 2002) when counting from projection images (0.6 spines/µm in *db331-GAL4; UAS- mCD8-GFP* flies).

Intrigued by the morphological resemblance of these protrusions to dendritic spines it was decided to address whether they bear additional characteristics of dendritic spines, such as actin enrichment and the ability to establish excitatory synaptic contacts.

3.4 LPTC spines are enriched in actin

To examine the cytoskeletal organization of LPTC neurons in adult animals GFP-tagged actin (Verkhusha et al., 1999) (Figure 3.3F-I) or tubulin (Grieder et al., 2000) (Figure 3.3K-N) were expressed using *db331-GAL4* and compared to the localization of membrane-tagged GFP (mCD8-GFP; Figure 3.3A-D). In all cases cytoplasmic mRFP was coexpressed (red in Figure 3.3A, C, E, F, H, J, K, M, O). Actin strongly accumulated in the spines, showing a clear enrichment in comparison to the dendritic branches (Figure 3.3H, I, arrows). The actin-GFP signal was quantified in spines and dendrites, normalized to the cytoplasmic mRFP (Figure 3.3E, J, O), and determined to be 2.5 to 8 times higher in spines than in dendrites. A similar localization was observed upon overexpression of GMA, a GFP-tagged version of the actin-binding fragment of moesin (Edwards et al., 1997), which is a faithful reporter of actin organization (Dutta et al., 2002). In contrast, tubulin was mainly localized in the primary and secondary branches, and proportionally less abundant in high order branches (Figure 3.3K-N). In fact, the tubulin-GFP signal intensity, normalized to the cytoplasmic mRFP, was 3.6 to 8 times lower in spines than in the dendrite shaft (Figure 3.3M-O).

To rule out an effect of actin or GMA overexpression in these neurons on spine morphology and density, spines of LPTCs overexpressing actin-GFP or GMA were compared with spines of LPTCs overexpressing mCD8-GFP. In each case a membrane-tagged RFP, myr-mRFP was

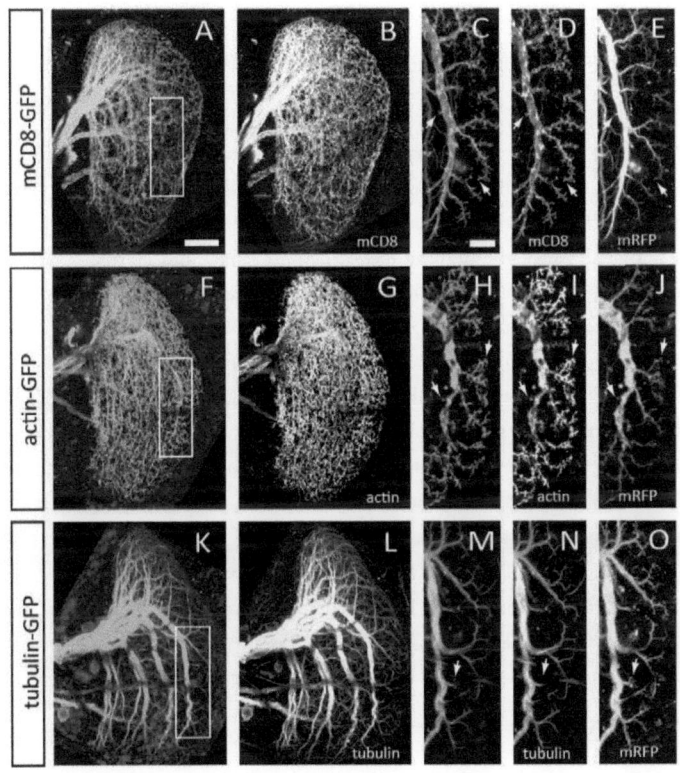

Figure 3.3 | ***Drosophila*** **Lobula Plate Tangential Cells have spines that are enriched in actin**

LPTCs expressing membrane tagged GFP (*mCD8-GFP*, A-D, green in A and C), *actin-GFP* (F-I, green in F and H) or *tubulin-GFP* (K-N, green in K and M) under the control of the *db331-GAL4* driver, together with cytoplasmic *mRFP* to visualize the morphology of the neurons (E, J, O and red in A, C, F, H, K and M). The colour panels show a merge of the mRFP signal (red) and the respective GFP-tagged construct (green). Note the spines (arrows in C-E, H-J and M-O), the high enrichment of actin in spines (arrows H-J), and the absence of detectable tubulin in those structures (arrows in M-O). Scale bars: 20 μm (A, B, F, G, K, L) or 10 μm (C-E, H-J, M-O). All images in this figure were taken by Ewa Koper.

coexpressed, to visualize spines appropriately and to quantify both spine length and density in the red channel (myr-mRFP). The density or the length of dendrite spines were not modified upon overexpression of *actin-GFP* or *GMA* in these neurons. Therefore, LPTCs possess spines that are enriched in actin and devoid of tubulin.

3.5 LPTC spines fall into four morphological categories

Although spines can modify their morphology even in the adult brain (Lendvai et al., 2000; Majewska and Sur, 2003), they have been traditionally divided in morphological categories that seem to reflect their level of maturation and of activity (Fiala and Harris, 2001; Nagerl et al., 2004; Noguchi et al., 2005). The morphology of the LPTC spines is also quite varied Figure 3.4A, B). In projections of optical section stacks the vast majority of LPTC spines were found to fall into four categories (Figure 3.4C). These are: *stubby* spines (44.4 ± 4.7%), if the diameter of the neck is similar to or greater than the total length of the spine; *thin* spines (24.9 ± 3.9%), if the length is greater than the neck diameter, and the diameters of the head and the neck are similar; *branched* spines (16.1 ± 3.1%), spines with up to three heads from a single neck; and *mushroom* spines (14.5 ± 3.9%), if the diameter of the head is greater than the diameter of the neck. A small proportion of protrusions (approx. 10% of all protrusions regardless of their length) was longer than 3μm. These structures were classified as filopodia.

Thus, LPTC spines fall into previously described spine categories. Similarly to what has been reported in other systems (Harris et al., 1992), the percentage of the possibly mature, mushroom spines is approximately 15% of the total number of spines.

3.6 LPTC spines represent sites of synaptic input

Mature dendritic spines are sites of synaptic input (Nagerl et al., 2004). To address whether the observed processes are functional homologues of vertebrate spines it was investigated whether they also are sites of synaptic input. Immunohistochemistry on *in toto* brains of adult flies expressing actin-GFP specifically in LPTCs was performed using nc82 antibodies against the pre-synaptic marker Bruchpilot (Kittel et al., 2006; Wagh et al., 2006), followed by confocal microscopy optical sectioning (Figure 3.5). A punctate staining restricted to the neuropils was obtained (Figure 3.5A), which was absent in the negative controls (secondary antibody only, data not shown). Puncta representing pre-synaptic sites were present both along the length of primary and secondary branches (Figure 3.5B) and juxtaposed to the spines (single optical sections in Fig. 3C arrowheads).

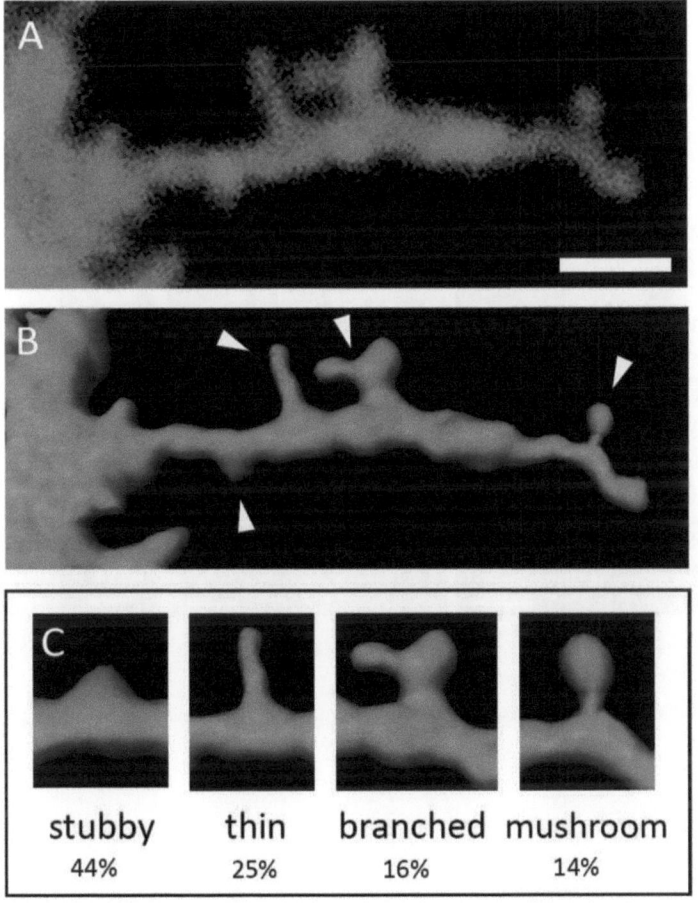

Figure 3.4 | **Classification of dendritic spines of LPTCs**

(A) A representative small dendritic branchlet of the medial region of VS1 showing the different types of dendritic spines. *UAS-mCD8-GFP* is expressed in LPTCs under the control of *db331-GAL4*. (B) Surface-rendered 3D reconstruction (Amira) of the same fragment. Selected spines are marked with arrowheads and magnified in the subsequent C panels. (C) Four categories of spines are detected: stubby, thin, branched, and mushroom, based on the spine length as well as the ratio between their maximum head and minimum neck diameter, as described in the text. The numbers in C represent the percentage of each spine category. >600 spines from 5 animals were analyzed. Scale bar: 2 µm.

To estimate the subset of spines in contact with a pre-synaptic terminal, the number of sites of close proximity (<0.1μm) between mCD8-GFP and the nc82-positive puncta in single confocal sections (Figure 3.5C) was counted. Additionally, 3D reconstructions of dendrite tree fragments were generated and rotated: only the contact sites that were maintained at all angles of rotation were counted. Based on these data it was observed that 88% (± 5%) of the spines were juxtaposed to nc82-positive puncta. Thus, most LPTC dendritic spines appear to receive synaptic input.

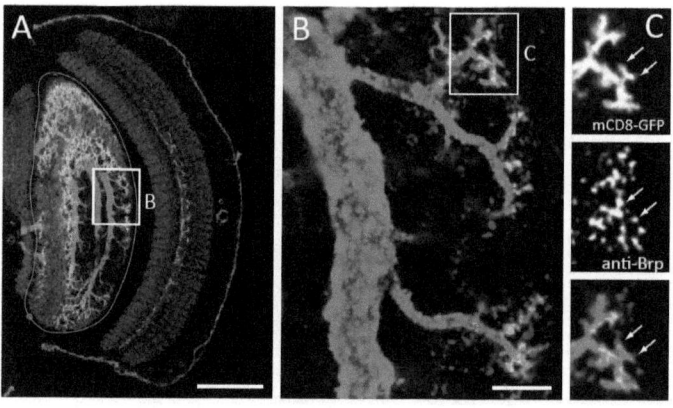

Figure 3.5 | **LPTC spines receive synaptic input**

(A) Single optical section through the LPTC dendritic field visualized by *db331-GAL4* driven expression of *mCD8-GFP* (green) and stained with anti-Bruchpilot antibody (nc82, red). Presynaptic puncta are highly abundant in the region containing the LPTC high order dendritic branches. The lobula plate is indicated. Scale bar: 50μm. (B) Single optical section through a representative fragment of VS1 (as indicated in A). Scale bar: 10μm. (C) Single channel images of a small dendritic fragment illustrating juxtaposition between spines and nc82 puncta. 88% of spines were in immediate proximity (<0.1 um) to presynaptic nc82 puncta as determined using 3D reconstructions.

3.7 Ultrastructure of LPTC spines

To provide additional lines of evidence that spines are sites of synaptic input, Ewa Koper analyzed the ultrastructure of spines in immuno-electronmicroscopy experiments that allowed unambiguously identifying the LPTCs in brain sections. LPTC spines were identified in serial brain sections: only the processes of up to 3μm in length that were included within 6 sections (50nm/section) were considered as spines. It was analyzed whether the spine would bear synaptic contacts. Indeed in all cases examined (n=5) T-bars (Prokop and Meinertzhagen, 2006), indicators of synaptic zones (Kittel et al., 2006; Wagh et al., 2006; Zhai and Bellen,

2004), were present immediately next to the spines. The T-bars were localized on the spine head in 4 examples, but they were also found on the neck and on the shaft of spines. In one example there were multiple T-bars on one spine. In two instances the presence of synaptic vesicles at a T-bar pre- synaptic to a spine could clearly be observed. These findings indicate that spines present on LPTCs can form active synapses.

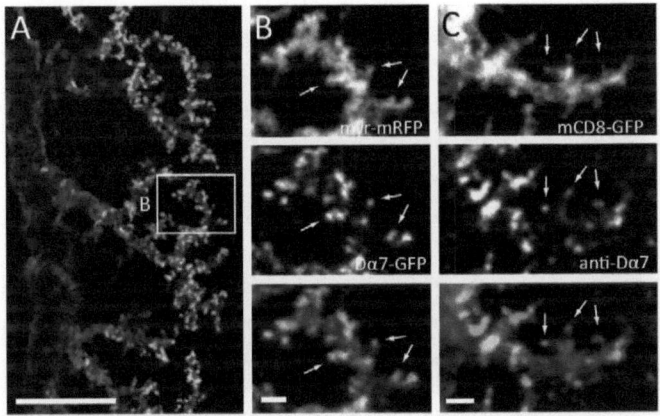

Figure 3.6 | Dα7 is localized at dendritic spines

(A) Projection of a confocal stack (spanning approx. 3 µm) through a representative fragment of VS1 expressing *myr-mRFP* (red) and *Dα7-GFP* (green) under the control of *db331-GAL4*. (B) Magnified view of the branchlet boxed in A, showing the single channels and the merge, as indicated. >90% of all spines contain the ectopically expressed ACh-receptor subunit Dα7-GFP. Note also the specificity of localization to the spines: 70% of the Dα7-GFP positive puncta localize to spines. (C) View of a branchlet of VS1, expressing *mCD8-GFP* under the control of *db331-GAL4* and immunolabelled with anti-Dα7 antibodies. The single channels and the merge are shown. Arrows point to single spines. Scale bars: 10 µm.

3.8   LPTC spines represent sites of excitatory synaptic input

Input onto vertebrate dendritic spines is mainly of excitatory nature (Gray, 1959). Previous immunohistochemical work demonstrated that LPTCs are likely to receive cholinergic synaptic input and acetylcholine is known to be the major excitatory neurotransmitter in the fly CNS (Brotz et al., 1995; Brotz et al., 2001). It was asked whether LPTC dendritic spines receive cholinergic input. To this aim, GFP-tagged versions of the acetylcholine receptor subunit Dα7 (Grauso et al., 2002) were generated. Upon coexpression of this construct and a membrane-tagged red fluorescent protein (myr-mRFP) under the control of *db331-GAL4*, the puncta were found to be primarily localized to dendritic spines (Figure 3.6A). It was determined that the vast majority of spines (97.7%, STDV 2.7%, >250 spines from 5 animals counted) contain Dα7-GFP puncta. Furthermore, 70% of Dα7-GFP puncta localized to spines (STDV 5%, >500 Dα7-GFP puncta counted), indicating that spines represent a preferential localization for Dα7-GFP. Importantly, the expression of this construct does not alter spine density (1.09 spines/µm versus 1.15 spines/µm in the control, ttest 0.3), indicating that it does not induce the formation of ectopic spines. Additionally, the Dα7-GFP puncta are significantly more often juxtaposed to presynaptic puncta positive for the nc82 antibody than expected for random distribution, suggesting that at least a subset of those represent sites of synaptic connectivity. To address if the ectopically expressed Dα7-GFP reflects the localization of the endogenous protein, anti-Dα7 antibodies were used in immunohistochemistry and a strikingly similar distribution of Dα7-positve puncta was obtained (Figure 3.6C). Taken together, the localization of Dα7 indicates that at least a vast majority of the synaptic input onto LPTC spines is excitatory.

In conclusion, the spines observed on LPTCs are morphologically similar to vertebrate spines; they are enriched in actin, devoid of tubulin and are sites of synaptic input, which is mostly excitatory. I thus conclude that these processes indeed represent spines.

3.9 dRac1 overexpression alters LPTC spine density and morphology

To test whether dendritic spines could be genetically modifiable in this system, the level of a factor known to affect spine morphology and density in vertebrates were manipulated. *dRac1* was chosen because the effects of Rac1 on spines in vertebrates are particularly well characterized (Govek et al., 2005). Full length *dRac1* (FL: rac1.L; (Luo et al., 1996; Nakayama et al., 2000) as well as dominant negative (DN: rac1.N17; (Luo et al., 1994) and constitutively active (CA: rac1.V12; (Luo et al., 1994) versions of *dRac1* were expressed specifically in the LPTCs under the control of *db331-GAL4*, and coexpressed *GMA* was used to visualize the dendritic trees. Since overexpression of CA *dRac1* led to lethality at pupal stages only the effects of full length and dominant negative *dRac1* over-expression could be analyzed (Figure 3.7). The overall dendritic architecture in both genotypes appeared to be similar to the wild type condition: neither position nor branching patterns of primary and secondary order dendrites were obviously affected. This is consistent with previous evidence that alteration of Rac activity does not affect the dendrite structure of pyramidal neurons or of cerebellar Purkinje neurons (Luo et al., 1996; Nakayama et al., 2000). However, a difference in spine morphology and an increase in spine density was noticed (compare Figure 3.7B and Figure 3.7A). To quantify these observations spine density analysis was quantified as described above (Figure 3.7G). Spine density was found to be increased by around 30% upon overexpression of either full length *dRac1* (Figure 3.7G; 1.54 spines/µm; n= 5; p= 0.0064 by t-test) or dominant negative *dRac1*

(1.51 spines/μm; n=5; p= 0.0051), in comparison to the control expressing *myr-mRFP* (1.15 spines/μm; n=5). Moreover, spines appeared shorter and less well defined. Although opposite effects might be in principle expected upon overexpression of full length and dominant negative proteins, both genotypes appeared indistinguishable and yielded similar results in the quantification (see Discussion). Next, it was addressed whether the processes present upon Rac1 over-expression share characteristics of spines as described above. Actin and Dα7 distribution were therefore analyzed (Figure 3.7B, C, E, F). Actin was found to be enriched in the Rac1-induced spines 2–7 times more than in the dendrite shaft in comparison to cytoplasmic mRFP (VS1 fragments from 5 animals quantified; Figure 3.7B, E). Dα7-GFP was also present on these processes (Figure 3.7F). Because of low signal level in these experiments, reliable quantifications could not be done. Rac1-induced spines seem thus to share two distinct characteristics of LPTC spines. Taken together, alterations of Rac1 levels appear to be capable of modulating spine density in *Drosophila* as previously reported for vertebrates.

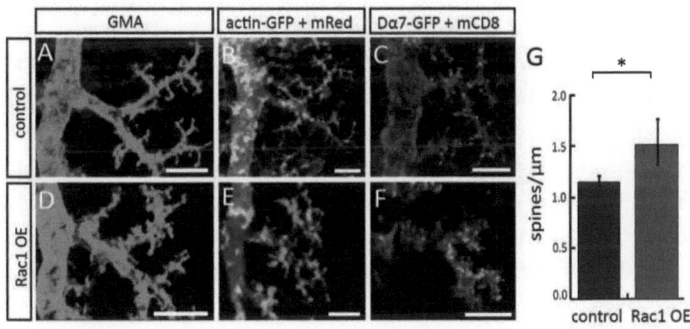

Figure 3.7 | **Spine density is modulated by Rac1**

(A-F) Similar dendritic fragments of VS1 are shown for animals carrying *db331-GAL4* and that were heterozygous for *UAS-GMA* (used for imaging) and either *UAS-myr-mRFP* (control, A) or *UAS-rac1.L* (D); *UAS-actin-GFP and UAS-mRed* with (E) or without (B) *UAS-rac1.L* or *UAS-Dα7-GFP* and *UAS-mCD8-cherry* with (F) or without (C) *UAS-rac1.L* respectively. (G) Quantification of spine density. >500 spines from 5 animals (as shown in A and G) were analyzed per data point. T-test p< 0.007. Scale bars: 5 μm.

In summary, Ewa Koper and I characterized the morphology and cytoskeletal organization of Lobula Plate Tangential cell (LPTC) dendrites and demonstrate that they bear small protrusions that closely resemble dendritic spines of vertebrates. LPTCs were considered to be an interesting model system to study dendrite and spine morphogenesis genetically. The first step towards this goal was the identification of Rac1 as a modulator of spine morphology in flies as previously described for vertebrates. Rac1 was identified as an interesting candidate molecule in initial attempts to screen for factors involved in dendritogenesis. These experiments will be introduced in more detail after the presentation of anatomical work describing Kenyon cell dendrites and their synaptic partners because the attempts to identify genes involved in dendrite morphogenesis were based on the anatomical knowledge of both LPTCs and Kenyon cells.

3.10 Microglomerular complexes in the mushroom body calyx

Kenyon cells are no ideal cellular system to study dendritic morphology because of their poorly defined morphology and the lack of a GAL4 line allowing the labelling of individual Kenyon cells. The morphology of single Kenyon cells can only be studied using MARCM (Lee and Luo, 1999). Kenyon cell dendrites form several branches throughout the calyx which often end with characteristic claw-like structures. Because of the high number of Kenyon cells (around 2000 per hemisphere) and the lack of specific markers it is an open question if they have an individually stereotyped morphology – but electrophysiology combined with theoretic considerations suggest that this might not be the case (Murthy et al., 2008). Although Kenyon cell dendrites seem to share some common features they are not very stereotyped between different cells. These constraints make cellular studies with the resolution presented for the LPTCs impossible. A number of advantages seemed to outweigh these technical shortcomings. First, Kenyon cells are known to play an important role in learning and memory. Second, Kenyon cells dendrites are the intrinsic elements of the mushroom body calyx and it has been shown that the volume of the calyx is sensitive to alterations of sensory environment in a number of insect species including the fly. Third, they are part of the well-characterized olfactory system. It is possible to manipulate the activity of the presynaptic partners of Kenyon cells with both odour stimulation and molecular tools. While the first two points make it plausible that Kenyon cell dendrites might undergo morphological rearrangements in an activity-dependent manner the third opens interesting opportunities for ultimately investigating the relation between neuronal activity and structural plasticity. For these reasons I decided to investigate if the mushroom body calyx (with the Kenyon cells as the intrinsic elements) could be a suitable model system for structural plasticity in the fly. I will describe the anatomy of microglomerular complexes in the mushroom body calyx before presenting my attempts to exploit this anatomical information for studies on structural plasticity.

3.11 Actin-rich microglomerular structures are present throughout the calyx

Since neuronal plasticity might require actin-dependent changes in dendrite morphology I looked for subcellular sites of actin accumulation in the adult calyx.

I expressed actin-GFP (Verkhusha et al., 1999) in most Kenyon cells using the *ok107-GAL4* driver (Connolly et al., 1996) and found ring-like calycal structures highly reminiscent of microglomeruli previously described in other insects based on phalloidin labelling (Frambach et al., 2004; Groh et al., 2004). In single confocal sections, microglomeruli were identified by a ring of actin-GFP fluorescence (diameter 1.75µm ± 0.30µm, n>200) surrounding a centre which was devoid of signal (diameter 0.66µm ± 0.14µm, n>200). To demonstrate the actin-enrichment in the ring-like structures I coexpressed actin-GFP and cytoplasmic mRFP using the *ok107-GAL4* driver (Figure 3.8B). Whereas mRFP was predominantly seen in the cell bodies, actin-GFP was specifically localized in the ring-like structures and was barely detectable in the cell bodies. Similarly to what has been reported in other insects, microglomeruli can also be visualized with phalloidin (Figure 3.8C). I conclude that actin is selectively enriched postsynaptically within microglomerular structures in the calyx.

3.12 Claw-like endings from different Kenyon cells constitute microglomeruli

Microglomeruli appear very similar upon phalloidin labelling (Figure 3.9C) or ectopic expression of actin-GFP in Kenyon cells (Figure 3.9B) suggesting a dominant contribution from Kenyon cells to the actin-enriched structures. To reveal the contribution of individual Kenyon cells I used the MARCM technique (Lee and Luo, 1999). Most Kenyon cells form few dendritic branches which usually do not bear many higher order branches and terminate with claw-like endings (Lee et al., 1999; Zhu et al., 2005) or similar structures (Strausfeld et al., 2003). Since I did not attempt to resolve the subtle morphological differences between the dendritic endings of Kenyon cells previously described using Golgi impregnations (Strausfeld et al., 2003) I decided for simplicity to subsume the Kenyon cell dendritic endings with the term 'claw-like endings' for the purpose of this study. I obtained high-magnification images of dendrites of single Kenyon cells expressing mCD8-GFP (Lee and Luo, 1999) and identified small processes protruding from the claw-like endings and from the dendrite branches (Figure 3.9C, D arrowheads), which were between 0.3 and 1 µm long (occasionally longer along the dendrite branches). These processes often resembled vertebrate spines in their morphology (e.g. mushroom shaped spines with a spine head diameter exceeding the spine neck diameter, see Figure 3.9D arrowheads), while some resembled filopodia or were of irregular shape. In contrast to the specific enrichment of actin in dendritic spines reported in vertebrates and in *Drosophila* (see above) actin appeared to be enriched in the entire claw-like ending, including the small spine-like protrusions, as shown in single confocal sections (Figure 3.9E-G). While claw-like endings displayed many spine-like protrusions (Figure 3.9D) these were found only sporadically along the dendritic branches (Figure 3.9C, arrowhead).

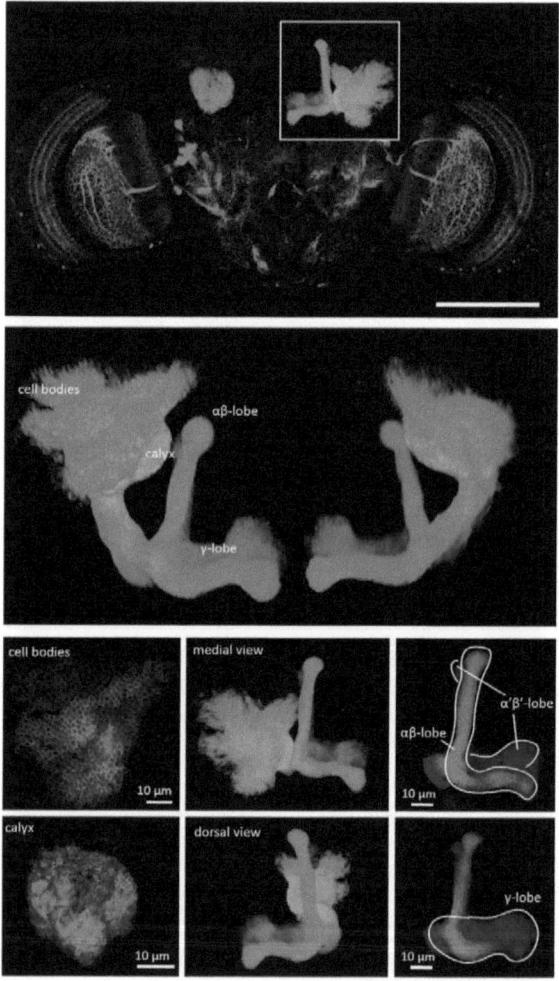

Figure 3.8 | **Mushroom body overview**

Single confocal section illustrating the position of the mushroom body calyx (upper panel, arrow). A 3D reconstruction of the right side of the mushroom body is included to illustrate the 3D orientation; magnified and rotated versions are presented in medial panels. Single confocal sections of the Kenyon cell body area, the medial calyx (bottom left panels), αβ, α'β'- and γ-lobes (bottom right panels). Scale bars: 150 μm (upper panel) or 10 μm.

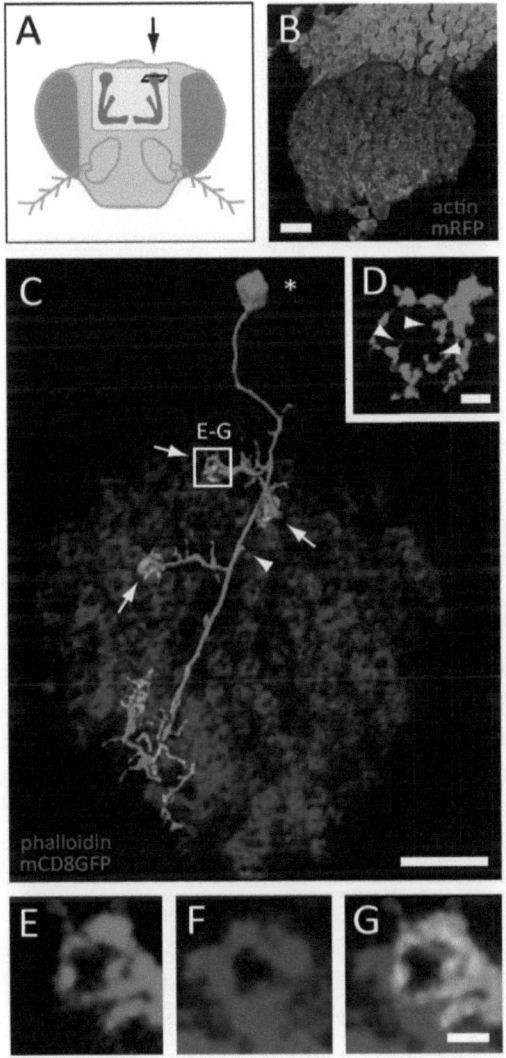

Figure 3.9 | **Actin-enriched microglomeruli in the mushroom body calyx are constituted by claw-like endings from Kenyon cell dendrites**

A) Scheme of the fly head to illustrate plane and orientation of all subsequent images. B) Single confocal section through the mushroom body calyx expressing actin-GFP (red) and cytoplasmic mRFP (green) in Kenyon cells, driven by ok107-GAL4. In contrast to mRFP, actin is selectively enriched in microglomerular substructures in the calyx as compared to the cell bodies. Scale bar: 10μm. C) Kenyon cell MARCM clone, labelled with mCD8-GFP (green) and shown as a 3D reconstruction projected onto a single plane of phalloidin labelled microglomeruli (red). Claw-like endings of Kenyon cell dendrites are indicated by arrows, the cell body by an asterisk. Scale bar: 10μm. D) Claw-like ending similar to the boxed region in C). Scale bar 1 μm. Spine-like protrusions are marked with arrowheads. E-G) Magnification of the boxed region in C) demonstrating that microglomeruli contain Kenyon cell claw-like endings. Scale bar: 1μm.

Due to the morphological resemblance and similar dimensions of microglomeruli and claw-like endings of Kenyon cell dendrites I suspected each microglomerulus to be constituted by claw-like endings from several Kenyon cells. Indeed, claw-like endings of single cell MARCM clones co-localized with phalloidin labelled microglomeruli (n=5) (Figure 3.9E-G). A single claw-like ending never entirely covered an actin-rich ring: I thus conclude that several claw-like endings from different Kenyon cells (see below) collectively form a microglomerulus. Each Kenyon cell has several claw-like endings and thus contributes to several microglomeruli (Figure 3.9C).

3.13 Microglomeruli are sites of synaptic contact

To investigate the location of synapses relative to the actin-enriched rings formed by the Kenyon cell claw-like endings I labelled synapses with the presynaptic markers synaptotagmin or Bruchpilot (nc82) while visualizing the microglomeruli using actin-GFP driven by *ok107*-GAL4 (Figure 3.10A, B). I found most presynaptic puncta to be located within the actin-devoid centres of microglomeruli; often outlining their inner edges (Figure 3.10B). Since there were many presynaptic puncta in the calyx, I wondered whether this localization would just be due to random distribution, but the percentage of puncta within the microglomerular centres turned out to be significantly higher ($52\pm11\%$, 2950 puncta counted; ttest $<4 \times 10-8$) than expected for random distribution ($22\pm5\%$). In single cell MARCM clones I could identify multiple presynaptic puncta outlining the inner edge of claw-like structures on single optical sections (Figure 3.10C-F) demonstrating juxtaposition. Projection neurons represent the only cholinergic neurons in the mushroom body calyx (Yasuyama et al., 2002). Using anti-ChAT labelling I confirm that the majority of the presynaptic boutons within microglomeruli are cholinergic (Figure 3.10G, H) and thus likely represent projection neuron boutons.

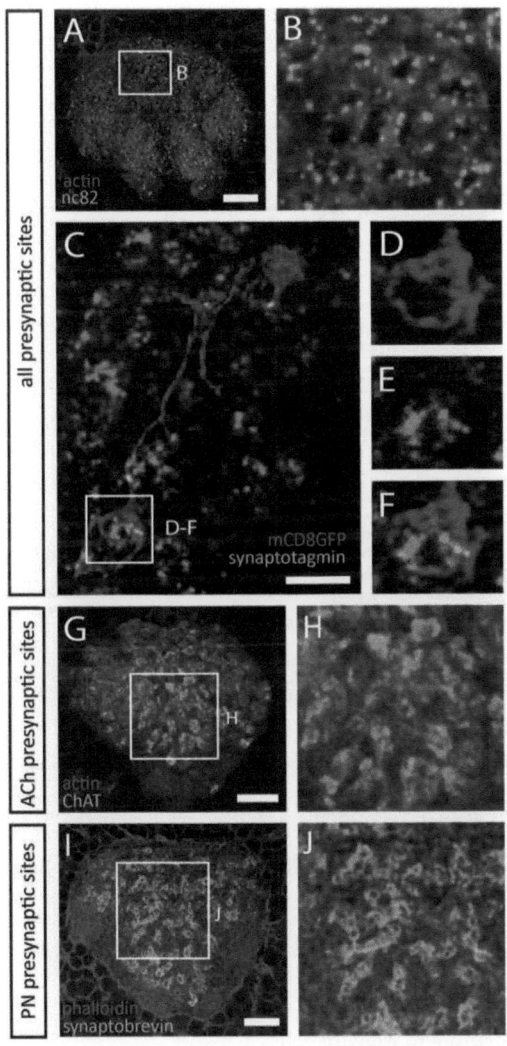

Figure 3.10 | **Synaptic organization of calycal microglomeruli**

A) Single confocal section through the mushroom body calyx expressing actin-GFP (red) in Kenyon cells; nc82 (green) labelling reveals presynaptic sites. Scale bar: 10μm. B) Magnification of the boxed region in A); presynaptic sites are outlining the inner rim of microglomeruli. C) Kenyon cell MARCM clone, labelled with mCD8-GFP (red), co-labelled with α-synaptotagmin (green). Scale bar: 5 μm. D-F) Magnification of the boxed region in C). A single optical section is shown. G) Single confocal section through the mushroom body calyx expressing actin-GFP (red) in Kenyon cells; α-ChAT (green) labelling reveals presynaptic cholinergic sites. Scale bar: 10 μm. H) Magnification of the boxed region in G). I) Single confocal section through the mushroom body calyx expressing synaptobrevin-GFP (green) in projection neurons; microglomeruli are colabelled with phalloidin (red). Synaptobrevin-GFP reveals spheroid presynaptic specializations of projection neurons. J) Magnification of the boxed region in I); projection neuron presynaptic boutons constitute the centres of microglomeruli and contact surrounding Kenyon cell dendrites.

3.14 Projection neurons constitute the centres of microglomeruli

Olfactory projection neurons send their axons from the antennal lobes to the mushroom body calyces where they form synapses with Kenyon cell dendrites (Stocker et al., 1990). Studies in other insects have demonstrated that dye-filled projection neurons constitute the centres of actin-rich microglomeruli (Frambach et al., 2004; Groh et al., 2004), suggesting a similar organization also in *Drosophila*. I labelled projection neurons genetically using the *gh146-GAL4* driver (Stocker et al., 1997) and the reporters mCD8-GFP (data not shown) or synaptobrevin-GFP (Ito et al., 1998) (Figure 3.10I, J), a presynaptic marker. Co-labelling with phalloidin revealed that the centres of the actin-rich microglomeruli are indeed defined by projection neuron presynaptic specializations (Figure 3.10I, J). Spheroid projection neuron boutons could be identified throughout the calyx and their dimensions matched those of the actin-devoid centres of microglomeruli, suggesting that a single projection neuron bouton constitutes the centre of a microglomerulus. In single confocal sections synaptobrevin-GFP-labelled projection neuron boutons appeared as ring-like structures (as also previously shown by (Ashraf et al., 2006)) similar to but slightly smaller than the actin-rich rings formed by the Kenyon cell dendrite endings. Single projection neuron boutons were surrounded by phalloidin labelled rings and outlined their inner edge and thus were highly reminiscent of the anti-ChAT labelling (compare Figure 3.10I, J and Figure 3.10G, H).

3.15 Acetylcholine receptors in Kenyon cells

Since actin-rich protrusions from Kenyon cells surround cholinergic projection neuron boutons I asked if they bear acetylcholine receptors. I expressed a GFP version of the Dα7 subunit of the receptor and found a striking localization at the inner rim of the actin-rich rings. Costaining with nc82 revealed that Dα7 receptor patches are often in very close proximity of presynaptic staining. As the projection neuron to Kenyon cell synapses are the only known cholinergic synapses in the calyx the Dα7 receptor and the complementary ChAT staining appear well suited to specifically visualize this type of synaptic connection.

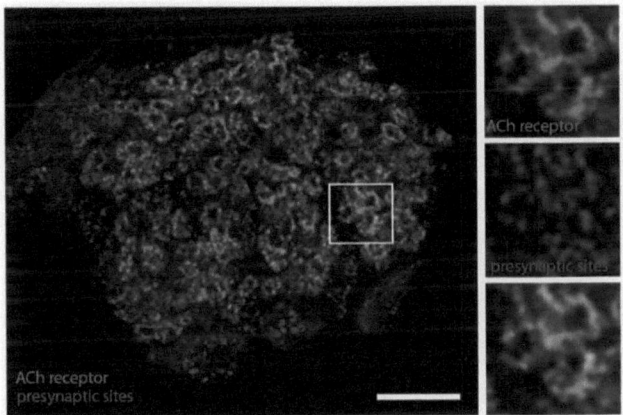

Figure 3.11 | **Acetylcholine receptors in Kenyon cells**

Medial, single confocal section of the mushroom body calyx expressing Dα7-GFP using the *ok107-GAL4* driver (green). Presynaptic sites are labelled with nc82 immunohistochemistry (red). The boxed region is magnified and individual channels are shown (right panels). Dα7-GFP localizes specifically to ring-like structures and often is in close proximity of presynaptic sites. Scale bar: 10 µm.

3.16 Microglomeruli are innervated by GABAergic interneurons

The mushroom body calyx in *Drosophila* as in other insects is innervated by GABAergic interneurons that form synapses with both Kenyon cells and projection neurons within microglomeruli (Ganeshina and Menzel, 2001; Yasuyama et al., 2002). To analyze how many microglomeruli receive GABAergic processes I expressed mCD8-GFP using the *gad1-GAL4* driver that labels GABAergic neurons (Ng et al., 2002). It is important to note that among the neurons labelled by the *gad1-GAL4* driver in the adult brain some non-GABAergic neurons and few Kenyon cells are also apparently included (based on position and size of the cell bodies). I found *gad1*-positive profiles to densely innervate the calyx and used co-labelling with phalloidin to count the percentage of microglomeruli with contribution from *gad1*-positive processes. I thus estimate that >90% of all microglomeruli contain *gad1*-positive processes (98%, STDV 1, n=3 brains; Figure 3.12A, B).

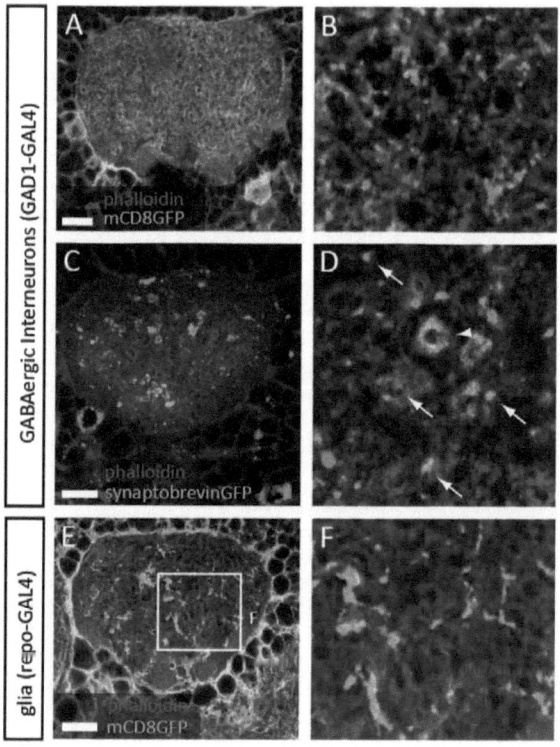

Figure 3.12 | **GABAergic interneurons and glial cells are present in the calyx**

A) Single confocal section through the mushroom body calyx expressing mCD8-GFP (green) in GABAergic interneurons (driven by *gad1-GAL4*); microglomeruli are co-labelled with phalloidin in this and subsequent panels (red). Scale bar in all panels: 10 μm. B) Magnification of the boxed region in A); processes from GABAergic neurons are extending essentially throughout the entire calyx. C) Single confocal section through the mushroom body calyx expressing synaptobrevin-GFP (green) in GABAergic interneurons (driven by *gad1-GAL4*). D) Magnification of the boxed region in C); discrete synaptobrevin-GFP puncta can be observed in many microglomeruli (arrows). Some microglomeruli seem completely labelled (arrowheads). E) Single confocal section through the mushroom body calyx expressing mCD8-GFP (green) in glia cells (driven by *repo-GAL4*). F) Magnification of the boxed region in E); glial processes appear not to contact all microglomeruli and seem not to define calycal subcompartments.

To identify possible sites of synaptic connectivity, I expressed Synaptobrevin-GFP using the *gad1-GAL4* driver and observed small Synaptobrevin-GFP dots often at or close to the inner rim of the actin-rich ring throughout the calyx (Figure 3.12C, D arrows) without any obvious regionalization. Since Kenyon cells are believed to be exclusively postsynaptic throughout the calyx this suggests that at least a significant proportion of the *gad1*-positive processes are presynaptic processes from GABAergic interneurons. In few microglomeruli, though, the Synaptobrevin-GFP signal seemed to line the entire actin-rich ring (Figure 3.12D, arrowhead) either corresponding to large GABAergic terminals identified by electron microscopy (Yasuyama et al., 2002) or representing a subset of microglomeruli in which GABAergic innervation is very abundant.

3.17 Glial processes are protruding into the calyx

Since glial processes enwrap the glomeruli in the antennal lobe I asked if the calyx is also subdivided by glia. Using the *repo-GAL4* driver (Sepp et al., 2001) to restrict the expression of mCD8-GFP to glia cells, I found the calyx (Figure 3.12E), pedunculus and lobes to be enwrapped by glial processes. I also noticed glial processes throughout the calyx (Figure 3.12F). Using 3D reconstructions I revealed that these do not enwrap individual microglomeruli (as opposed to locust (Schurmann, 1974), but similar to bees (Ganeshina and Menzel, 2001)) or distinct groups of microglomeruli and do not obviously reflect the quadripartite developmental origin of the calyx. Instead, they seem to form a rather loose meshwork throughout the calyx. I cannot exclude at this point that more subtle subdivisions exist. The overall organization of microglomeruli, as emerging from the above data, is schematized in Figure 3.13.

3.18 Around 1000 microglomeruli are found in the calyx

To address whether the microglomeruli identified by the actin-rich rings are distributed evenly throughout the calyx or may cluster in morphologically distinct subdomains I obtained serial confocal sections spanning the calyx of flies expressing actin-GFP driven by *ok107-GAL4* and generated 3D models. I identified microglomeruli at all sections through the calyx, marked individual ones at the plane of their maximal (xy) extension in a confocal z stack and determined their positions in x and y. Plotting this positional information of all microglomeruli of an animal did not reveal obvious clusters of preferred microglomerular position. Rather, the microglomeruli seemed to be randomly distributed throughout the calyx (as in Figure 3.15D-F). I determined the number of microglomeruli for 24-h old wild-type flies and obtained an average value of 786 ±73 (mean ± STDV, n= 3 calyces) per calyx. Because of limitations of the quantification method. I expect the actual number to be slightly higher (200 projection neurons x 5 boutons per projection neuron giving about 1000 microglomeruli per calyx).

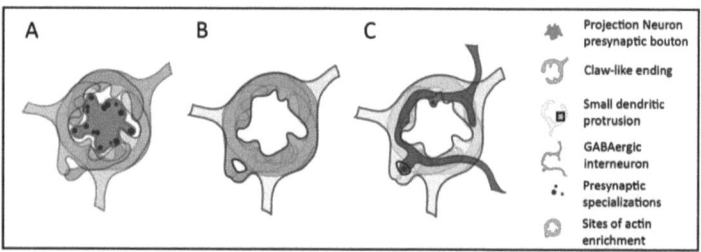

Figure 3.13 | **Schematic illustration of a microglomerulus**

A) Claw-like endings from several Kenyon cells (shades of green) enwrap and form synapses (red puncta: presynaptic sites) with a projection neuron presynaptic bouton (light red). B) Actin (yellow) is mainly localized to claw-like endings. C) Processes from GABAergic neurons (blue) contribute to and form presynaptic specializations (red) within microglomeruli. The precise localization of synapses is inferred from the literature (Yasuyama, 2002).

3.19 Microglomeruli have distinct presynaptic contributions

Projection neurons innervate specific antennal lobe glomeruli and thus carry distinct olfactory information. I have shown that microglomeruli receive excitatory cholinergic input from projection neuron boutons (Figure 3.10) and they can thus be classified according to the projection neurons that innervate them. To illustrate this I expressed mCD8GFP under the control of the *mz19-GAL4* (Ito et al., 1998) or *np3529-GAL4* drivers which label the projection neurons sending their dendrites to the antennal lobe glomeruli DA1, VA1d and DC3 or DL1 (Tanaka et al., 2004), respectively (Figure 3.14A, B). Consistent with recent observations (Lin et al., 2007) I found that each of these two projection neuron subsets reproducibly projects to a defined region in the calyx. Microglomeruli can thus be classified according to the projection neuron input they receive and hence possibly process distinct olfactory information.

3.20 Microglomeruli have distinct postsynaptic contributions

I then asked if microglomeruli can also be subdivided according to the Kenyon cell subtypes that constitute them. It has been suggested that the mushroom body lobes have distinct functions in learning and memory (Akalal et al., 2006; Krashes et al., 2007; McGuire et al., 2001; McGuire et al., 2003; Pascual and Preat, 2001; Zars et al., 2000). To restrict the expression of actin-GFP to different subsets I used the following Gal4-driver lines: *1471-GAL4* (Isabel et al., 2004), which drives expression in a large subset of γ-neurons and shows no expression in the vertical lobes; *201y-GAL4*, expressing in a large γ-subset and in some αβ-core neurons; *c739-GAL4*, which is specific for a large subset of αβ-neurons and *17d-GAL4*, specific for a smaller subset of αβ-neurons only (Yang et al., 1995). The total population of microglomeruli was labelled with phalloidin. I found the actin-GFP labelling to be restricted to distinct subpopulations of phalloidin-labelled microglomeruli in all four genotypes (Figure 3.15). In agreement with earlier reports (Strausfeld et al., 2003; Tanaka et al., 2004) I found predominantly γ-neuron dendrites in the centre (Figure 3.15G) and dendrites of αβ-neurons mostly in the periphery of the calyx (Figure 3.15A, H, I). Importantly, I found three different populations of microglomeruli in all three genotypes: microglomeruli without, with partial or dominant (Figure 3.15B) contribution of the genetically labelled subset.

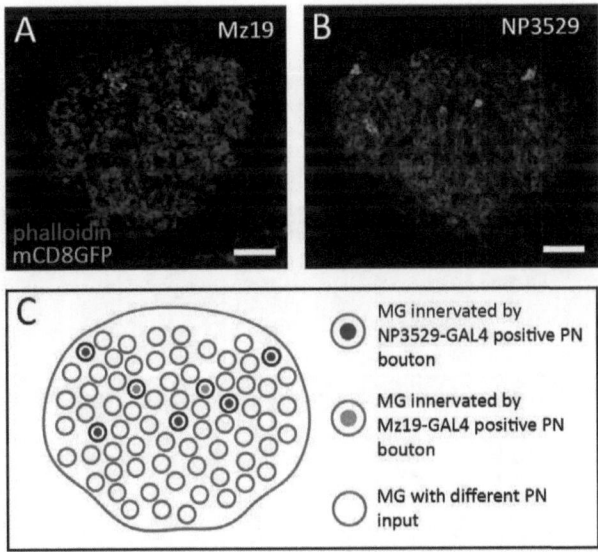

Figure 3.14 | **Microglomeruli differ in their presynaptic constituents**

Subsets of microglomeruli can be highlighted via mCD8-GFP labelling (green) of genetically defined subsets of presynaptic projection neurons as exemplified using the *mz19-GAL4* (A) or *np3529-GAL4* (B) drivers. Single confocal sections taken from a similar medial position are shown. Many more boutons were present on different sections in both cases. Phalloidin (red) reveals the total population of microglomeruli. Scale bars: 10μm. C) Scheme illustrating the differences in presynaptic constituents of microglomeruli.

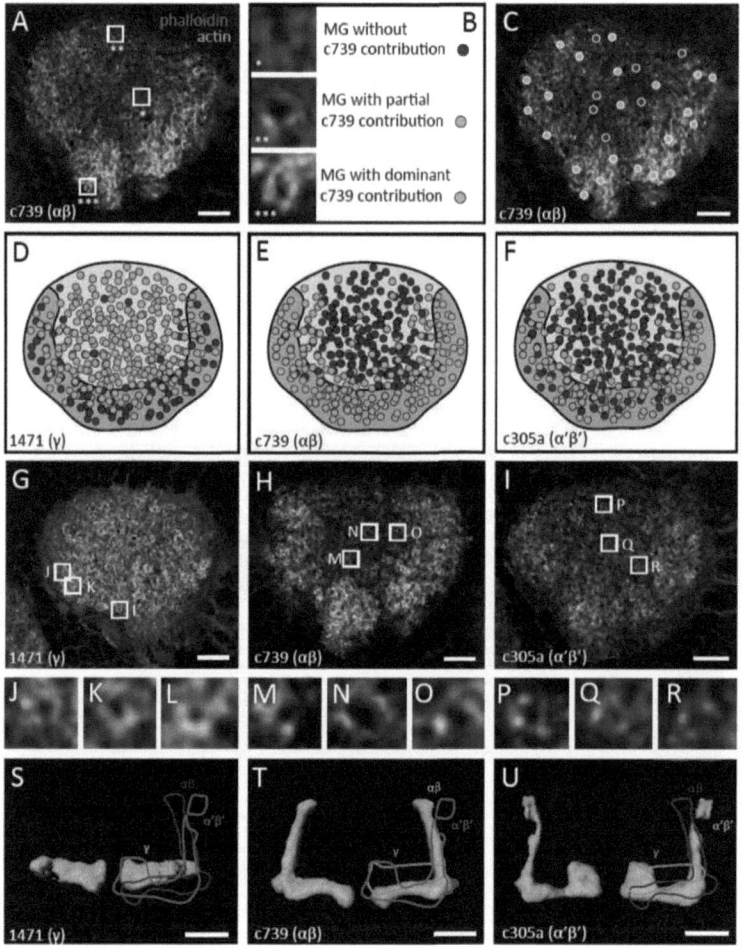

Figure 3.15 | **Microglomeruli differ in their postsynaptic constituents**

Subsets of microglomeruli were highlighted via actin-GFP (green) expression in γ-Kenyon cells (G, J-L, S; using *1471-GAL4*), αβ-Kenyon cell (A-C, H, M-O, T; using *c739-GAL4*) or α'β'-Kenyon cells (I, P-R, U; using *c305a-GAL4*), the entire population of microglomeruli was labelled by phalloidin (red). Microglomeruli without, with partial and with dominant contribution from the genetically labelled subset were identified in all three genotypes as exemplified in B). D-F) Illustrations of microglomerular distributions as categorized in B) using *1471-GAL4* (D), *c739-GAL4* (E) or *c305a-GAL4* (F) drivers.

Many microglomeruli in the central calycal area appear to be dominated by dendritic endings from γ-neurons (light grey in D-F), while most microglomeruli in the medial periphery seem to be dominated by αβ-neurons (dark grey in D-F). There were microglomeruli with partial contribution from γ-neurons in the αβ/α'β' -dominated region (J-L) and vice versa (M-R). J-R) Magnifications of selected microglomeruli in the boxed regions in G-I. S-U) 3D reconstructions of actin-GFP expression in the mushroom body lobes using *1471-GAL4* (S), *c739-GAL4* (T) or *c305a-GAL4* (U) drivers. Scale bars: 10μm.

These observations suggest that microglomeruli are not always formed by a single, defined type of Kenyon cell. Since I could not differentially co-label distinct Kenyon cell populations, I indirectly addressed this possibility by estimating the percentages of microglomeruli that different Kenyon cell subtypes contribute to. I estimate the *1471-GAL4* positive (γ) Kenyon cells to contribute to 70-90% of all microglomeruli, the *c739*-positive (αβ) to 30-60% and the *c305a*-positive (α'β') to 20-40% (see Figure 3.15D-F for illustrations).

Since these drivers define non-overlapping Kenyon cell subsets (Figure 3.15S-U) this suggests that at least 20% of all microglomeruli receive contributions from different Kenyon cell populations. To further strengthen this observation I searched for *1471*-positive (γ) microglomeruli within the main αβ-region (Figure 3.15J-L) or *c305a*-positive (α'β') or *c739*-positive (αβ) microglomeruli within the main γ-region (Figure 3.15M-R, F). I found many such glomeruli and suggest that they are prevalently composed of αβ-neurons, with a minor γ contribution or of γ-neurons with a minor αβ/α'β' contribution, respectively. My data suggest that distinct classes of microglomeruli exist that are formed by the dendritic claw-like endings of either a single Kenyon cell population or of different Kenyon cell populations.

3.21 Structural plasticity

I have characterized anatomical details of the connectivity between projection neurons and Kenyon cells in microglomerular structures. These synaptic complexes represent a repeated structural unit throughout the calyx and I asked if their morphological description allowed studying whether microglomeruli undergo experience-dependent alterations. Volumetric changes in the insect calyx have been reported upon alterations of sensory environment (Durst et al., 1994; Heisenberg et al., 1995; Withers et al., 1993). I asked if the structural plasticity demonstrated for the calyx could be partially due to rearrangements of the organization of the actin-rich calycal microglomeruli (Groh et al., 2006; Lent et al., 2007). To address this it would be ideal to study single Kenyon cells and their individual protrusions. Unfortunately, such an approach is hampered by the poorly stereotyped morphology of Kenyon cells. The morphology of microglomeruli, however, is roughly consistent between animals and thus ideal for addressing changes at a calyx-wide but cellular scale. In the following sections I will present my attempts to exploit the insights into the anatomy of calycal microglomeruli for studies on structural plasticity.

3.22 Technical challenges of studies on structural plasticity

Aiming at studying structural plasticity in the mushroom body calyx I had to meet a number of technical challenges. First, I had to develop reliable and precise tools for morphometric analysis. After using a manual approach initially I developed tools for automated image analysis to allow higher accuracy and consistency. Second, I had to generate genetic tools allowing the reliable visualization of microglomerular morphology in a GAL4/UAS independent way. Third, I needed a way to conclusively compare results obtained from animals of different size. I could then describe microglomerular morphology in the wild type and test the effects of a number of sensory manipulations. Finally, genetic tools allowing manipulating neuronal activities in defined ways were required to relate structural rearrangements with neuronal activity.

3.23 Manual quantifications of microglomerular morphology

I first measured the surface included within the actin-rich ring of the 25 largest microglomeruli (visualized using actin-GFP driven by *ok107-GAL4*) on single confocal sections cutting the calyx through the middle of the z axial extension. To address whether microglomerular morphology might be plastic I compared the average inner microglomerular surface of young animals (24 hours) to mature flies (14 d). I detected a significant (ttest <0.00035) increase in inner microglomerular diameter with age, from 0.29 μm^2 (±0.065, n=5) at 24 hours to 0.47 μm^2 (±0.1, n=5) at 14 d (Fig. 7) while the number of microglomeruli decreased by around 25% to approximately 750. These findings demonstrate that calycal circuitry undergoes morphological alterations during adult life. It seemed possible that the onset of sensory experience plays a crucial role in this context. To start addressing this possibility I deprived *UAS-actin-GFP/+; ok107-GAL4/+* flies of their olfactory afferents by cutting their third antennal segments and maxillary palps less than 12 hours after eclosion. While this was reported to kill the olfactory receptor neurons (Berdnik et al., 2006) projection neuron morphology appears unaffected (Berdnik et al., 2006; Tanaka et al., 2004). I analyzed microglomerular morphology as before and did not find significant alterations in inner microglomerular area upon sensory deprivation after 14 d (0.47 $\mu m2$ ±0.1, n=5, in control animals; 0.51 $\mu m2$ ±0.15, n=5, in deprived animals; ttest 0.44). However, the manual quantifications used in this set of experiments were not satisfactory and the results were thus considered inconclusive. The procedure to count microglomeruli suffered from the subjectivity of the criteria used for the identification of microglomeruli. Different criteria (such as shape and size) could yield significantly different numbers. The procedure to measure microglomerular area was not only very laborious but also had the following shortcomings: a) one calycal section does not allow accounting for intra-calycal variation and the position of the single section was not reproducible enough to allow conclusions on a specific region of the calyx; b) the total number of microglomeruli measured manually was possibly insufficient to be representative for the entire population; c) the criteria to choose the best-defined microglomeruli were subjective, non-explicit and not stable over time.

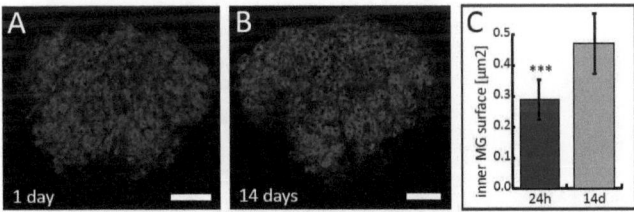

Figure 3.16 | **Microglomeruli rearrange during early adult life**

Medial calycal sections from one day (A) or 14 days (B) old animals expressing actin-GFP using the *ok107-GAL4* driver. Single confocal sections are shown. (C) The area of microglomerular centres was manually determined and significantly increased with age. Scale bars: 10 μm.

3.24 Automated image analysis

I therefore decided to develop software tools to automatically analyze images. Automated image analysis allowed increasing the number of calycal sections to get a realistic impression of the entire population of microglomeruli. The criteria used to detect microglomeruli were used consistently for the entire data set and were explicitly defined in the algorithm used for image analysis. Definiens software was used to develop a suitable algorithm allowing the detection of microglomerular rings and centres in a fully automated manner. A fully automated procedure was required to handle the large amount of data and to avoid subjective judgments during image processing.

3.25 Genetic tools used for imaging

Although published alterations in calycal volume upon manipulations of the sensory environment of flies suggested some degree of plasticity it was entirely unclear how these manipulations would affect microglomerular morphology (if at all) and to what extent.

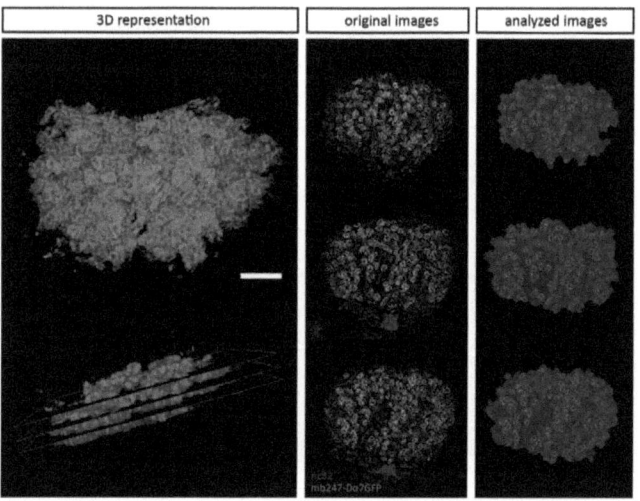

Figure 3.17 | **Overview of automated image analysis of the calyx**

A 3D reconstruction (Amira) of the calyx illustrating its shape and the high abundance of microglomerular structures (A). Confocal stacks of 50 sections spanning the calyx were used for morphometric analysis. Three such sections are illustrated and representative original images (*mb247-Dα7-GFP*, green; nc82 immunolabelling, red) are shown in (B). All images were analyzed fully automatically using Definiens software (C). Calycal volume was calculated as the sum of pixels assigned as calyx from all sections, microglomerular parameters were calculated as mean values from all objects identified per calyx.

It was thus crucial to develop tools to reliably measure as many different parameters of microglomerular morphology as possible and to aim at a precision that would allow detecting even subtle changes. At the same time image acquisition needed to be fast and consistent to allow taking large numbers of images. From all genetic labelling used for the anatomical description of the calycal connectivity the expression of Dα7-GFP in Kenyon cells appeared most suitable to meet these requirements. Automatic image analysis (see below) was facilitated by the facts that microglomerular rings were better defined than with actin-GFP and that the cell bodies were almost devoid of signal. Moreover, the localization of the ACh receptor could potentially reveal subtle alterations in synaptic organization that would be impossible to detect morphologically. To include information on the presynaptic components in the calyx the genetic labelling with Dα7-GFP labelling was combined with staining against the presynaptic marker Bruchpilot (nc82).

3.26 GAL4/UAS independent genetic labelling

Since I was aiming at genetic manipulations of neuronal activities exploiting available GAL4 driver lines to express molecular markers and tools in defined neuronal populations of the olfactory circuit I needed GAL4/UAS independent ways to label microglomerular morphology. The enhancer fragment of the *mb247-GAL4* driver, which includes a large number of (mostly αβ- and γ-neurons) Kenyon cells, was obtained from Dr. Andreas Thum and used to generate transgenic flies expressing either Dα7-GFP, actin-GFP, actin-KO or abp-KO in a large fraction of the Kenyon cells from direct fusion (GAL4/UAS independent) expression constructs.

3.27 Parameters determined with automated image analysis

The areas of microglomerular ring and lumen as well as shape parameters of both types of object were determined for every microglomerulus to quantitatively describe its morphology. Microglomerular rings were detected based on Dα7-GFP localization and microglomerular lumens (likely corresponding to presynaptic boutons) were detected as globular structures surrounded by Dα7-GFP signal.

Presynaptic puncta were detected and their area and average brightness were determined. Numbers of all object types were determined but were not used for analysis because the methodology was not considered adequate for precise counts. Calycal outlines were detected based on the borders of the Dα7-GFP signal and calycal volumes were determined. All these parameters were obtained from confocal stacks spanning the mushroom body calyx in 50 sections at a z resolution of around 0.25 μm. The volume of the central brain was determined from an independent set of 50 confocal images spanning the brain at a z resolution of around 2 μm.

3.28 Data normalization

Microglomerular morphology was determined from confocal images spanning the entire calyx and the physical sizes of objects were calculated based on the size of the pixels in each image. Unfortunately, insects vary considerably in body size, brain weight and brain volume. Since I could not track morphological changes upon sensory manipulations within the same brain – because I needed to dissect the brain for imaging – I needed means to compare animals within different experimental groups with each other. Absolute physical sizes of objects are not necessarily very good parameters for such a comparison if the size of the animals varies considerably. Previous studies have either underestimated this problem or used large numbers of animals per experimental group to deal with the amount of variation. In those studies, however, only relatively simple morphological parameters such as calycal volume have been assessed and it appeared unrealistic to determine parameters of microglomerular morphology on more than 50 brains per experimental group because determining these parameters requires many more images of higher quality.

For these reasons I decided to determine brain volumes and calycal volumes and tested if these parameters could be used to normalize size parameters of microglomeruli to account for differences in the size of the animals. Measurements of calycal volume appeared more accurate and consistent and were consequently chosen for normalization purposes.

3.29 Analysis of microglomerular morphology in the wild type

Using these tools I tried to quantitatively describe microglomerular morphology in the wild type. I obtained images from *mb247-Dα7-GFP* animals dissected within 2 hours after eclosion or aged for 2, 7 or 14 days. 7-15 animals per group and both calyces of all animals were analyzed. Differences of microglomerular size between different time points were detected but considered inconclusive due to considerable variation in brain volume of different animals. I normalized microglomerular size with calycal volume to compensate for size differences between animals. The ratio of average lumen (or ring) area/total calyx area rapidly increased within the first two days after eclosion and continued to increase at slower pace until 7 days before slightly decreasing again. This could either be due to an increase in microglomerular size or a decrease in calycal volume. Comparing calycal volumes to brain volumes (excluding optical lobes) I did not find any indication for decreasing calycal volumes. However, limitations in the methodology preclude precise statements on relative calycal volumes (calyx volume/brain volume).

The numbers of microglomeruli that were detected using automated image analysis increased with age. However, earlier manual analysis (see above) of the images suggested that the detection rate rather than the total number of microglomeruli increased. Visual investigation of the images indeed suggested that microglomerular rings were getting larger and better defined. They appeared more distinct and more clearly separated from each other and could be identified with less ambiguity. This likely resulted in higher recognition rates at older ages suggesting morphological refinement of microglomeruli.

I conclude that microglomerular complexes in the calyx undergo morphological rearrangements within the first days after eclosion.

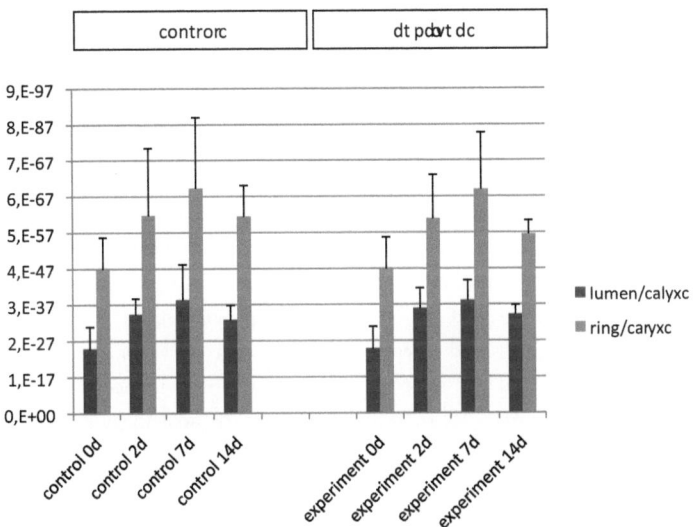

Figure 3.18 | **Microglomeruli rearrange during early adult life**

Results obtained with automated image analysis of calycal sections from both control animals (left) and from flies which had their antennae and maxillary palps removed within 2 hours after eclosion (right) are shown. Four different time points (0, 2, 7 and 14 days after eclosion) were obtained. The first data point in both cases is identical. The area of microglomerular rings (blue) and lumens (red) were determined and normalized with total calycal area (the summed calycal area from all sections). The normalized areas of ring and lumen increase with age before slightly decreasing again in the second week after eclosion. Similar results were obtained for both control and experimental (bilateral olfactory deprivation) groups. >5 animals were analyzed per data point, >1000 microglomeruli (in 2D, a single microglomerulus could be present on several sections) were identified and analyzed per animal.

3.30 Manipulations of sensory experience

Calycal volume has been reported to depend on rearing conditions, e.g. population density and sex of the partner fly (Heisenberg et al., 1995; Hitier et al., 1998). Rearing conditions also affect Kenyon Cell axonal branching pattern (Charles Tessier, personal communication). Moreover, manipulations of visual input demonstrated a dependence of calycal volume on vision (Barth and Heisenberg, 1997). I aimed at partially reproducing these published manipulations to explore if any of those reliably affects microglomerular morphology. Moreover, I decided to also include and to focus on olfactory deprivation because of the abundance of neuroanatomical knowledge on how odour molecules elicit neuronal responses in the olfactory pathway including the calyx. Moreover, it represents a single sensory modality (in contrast to e.g. social isolation).

To manipulate the flies' sensory environment the following paradigms were used: For bilateral or unilateral olfactory deprivation I surgically removed third antennal segments and maxillary palps on both or one side within 2 hours after eclosion; for bilateral visual deprivation flies were kept in vials wrapped in aluminium foil; unilateral visual deprivation was obtained by painting one eye with a black and the other with a transparent paint; for social isolation, flies were isolated as pupae and raised in eppendorf tubes; for starvation, flies were kept in vials supplying water but no food and sleep deprivation was done via mechanical agitation with a computer-controlled vortexer. All these manipulations of sensory environment were done starting within two hours after eclosion until flies were dissected at 48 h after eclosion (with the exception of starvation: flies were kept on vials with food for 24 hours and transferred to vials with water only for the remaining 24 hours before dissection). As for the wild type control I analyzed 7-15 animals per group and both calyces of all animals.

Alterations of microglomerular morphology were detected but remained difficult to interpret due to their rather subtle nature, limitations of the methodology and the relatively low number of animals. I decided to analyze olfactory deprivation in more detail and obtained additional data points at 7 and 14 days to increase the number of animals available for comparison with the control. The analysis of these data confirmed the rearrangements of microglomerular morphology during early adult life as described for the control and did not yield obvious differences. I conclude that the rearrangements during early adult life are large compared to possible alterations resulting from olfactory deprivation.

3.31 Genetic manipulations to alter activity in relevant neuronal populations

The experiments described above suffer from difficulties in relating morphological parameters of different animals with each other. To circumvent these problems it would be desirable to be able to compare two distinct populations of microglomeruli within the same brain. Suitable genetic tools were established. A description of these tools as well as the results obtained will be submitted as a primary research article and are therefore omitted here.

In summary, I developed tools for reliable genetic labelling of microglomerular complexes and for their quantitative morphological characterization using automated image analysis. Although methodological limitations remain I provide evidence for morphological rearrangements of these synaptic complexes during early adult life. Recent refinements of the method allow genetic manipulation of the synaptic input to a distinct population of microglomeruli and the comparison of their morphology to unaffected neighbouring microglomeruli

3.32 Genetic analysis of dendrite and spine morphogenesis

The experiments addressing structural plasticity were based on the morphological description of microglomerular complexes in the mushroom body calyx. This anatomical information could also be used for studies aiming at the identification of genetic factors contributing to dendrite morphogenesis. The opportunities of fly genetics in this regard were an important reason to start characterizing dendrites in the *Drosophila* CNS and the detailed anatomical description of both LPTCs and Kenyon cells provide a good starting point to investigate the feasibility of genetic approaches to dendrite differentiation in both systems. It is desirable to establish a system in the CNS that can be used to characterize dendritic development, to identify and characterize molecular players involved in the various aspects of this process because this would allow circumventing some of the limitations of similar approaches in the PNS. The *Drosophila* PNS has successfully been used to screen for genes involved in dendritogenesis but since the PNS dendrites do not receive synaptic input, an important aspect of neuronal development in the CNS, the establishment of synaptic contacts, cannot be studied in this system. In the following sections I will summarize my attempts to set up an experimental system to allow large scale genetic analysis of dendrite morphogenesis in the CNS.

3.33 LPTCs as an assay system for a genetic screen

The morphology of the LPTC dendritic field appears well suited to screen for genes involved in dendrite morphogenesis. Genes involved in dendritic branching and path finding might well be identified based on distorted morphology of LPTC dendrite field size, shape, organization or the localization and shape or individual primary dendrites. Finer morphological parameters such as the distribution and branching pattern of smaller dendritic branches, spine density and morphology can be obtained from high magnification images of the medial region of VS1. All of these parameters can be assessed by visual investigation within a couple of seconds at the confocal microscope (but quantitative investigations are challenging, see below). The main advantages of the system are the possibility to reliably look at a small group of individually identifiable neurons, their complex but stereotyped morphology and the presence of spines.

3.34 Calycal microglomeruli as an assay system for a genetic screen

The morphology of individual Kenyon cells is not very useful for genetic analysis of dendritogenesis due to the lack of stereotypy and GAL4 drivers with cellular resolution, but the morphology of calycal microglomeruli is roughly consistent between animals. Genetic labelling of microglomeruli allows looking at a population of around 2000 Kenyon cells which need to form dendritic branches with claw-like endings and form contacts with their presynaptic partners to establish the stereotyped appearance of microglomerular complexes. A problem in dendrite differentiation or synapse formation upon genetic manipulation could possibly be identified based on morphologically obvious alterations of microglomerular morphology.

3.35 GAL4/UAS based genetic manipulations

The investigation of candidate lines required dissection and visual screening at the confocal microscope and were thus relatively time-consuming. I therefore decided to apply cell specific manipulations of selected target genes using the GAL4/UAS system in order to increase the specificity and (hopefully) prevalence of effects (alternative approaches such as EMS or transposon mediated mutagenesis require screening of larger number of candidate lines). For a number of genes dominant-negative, constitutively active or overexpression UAS constructs are available and RNAi constructs under UAS control have been generated for >80% of the genome (Dietzl et al., 2007). The GAL4 drivers to specifically visualize the LPTCs and mushroom bodies, *db331-GAL4* (obtained from A. Borst) and *ok107-GAL4* (Connolly et al., 1996), respectively, can be used for genetic screens individually but they can also be combined to allow simultaneous visualization of both neuronal populations (see upper panels of Figure 3.2 and Figure 3.8). Most of the screening was done using the following stocks *db331-GAL4; UAS-actin-GFP; ok107-GAL4* or *db331-GAL4; UAS-actin-GFP; UAS-dicer2; ok107-GAL4* (see Material and Methods for details and alternative stocks).

3.36 Pilot screen I

To evaluate the feasibility of the approach and to optimize the screening procedure a pilot screen was done using a small set of available UAS constructs (including overexpression (OE), constitutively active (CA) and dominant negative (DN) variants) to manipulate a few candidate genes selected based on the literature.

Name	Source
fmr1.OE	Kendal Broadie, USA
sra1.DN/sra1.RNAi	Christian Klämbt, Germany
CaMKII.CA/CaMKII.DN/ala.OE	Leslie Griffith, USA
cdc42.CA/cdc42.DN	Bloomington stock centre
rac1.OE/rac1.CA/rac1.DN	Bloomington stock centre
rho1.CA/rho1.DN	Bloomington stock centre
trioGEF.CA	Bloomington stock centre
dia.OE	Jana Lindner, Tavosanis lab
dia.RNAi	Bloomington stock centre
creb.OE	Bloomington stock centre

From this set of experiments *rac1* was identified as a potential modulator of spine density and morphology and it was decided to characterize this gene in more detail (part 3.9). Overexpression of Diaphanous as well as manipulation of Cdc42 appeared to induce alterations in dendritic morphology, most notably of the primary dendrites. Manipulations of Rac1 and CaMKII led to morphological distortions of the mushroom body calyx. The following lines induced lethality prior to eclosion: UAS-rac1.V12 (CA), UAS-rho1.N19 (DN).

It was concluded that the screening procedure was suitable to detect interesting candidate genes for further analysis (at least Rac1) and to screen reasonably large numbers of UAS lines (up to 100 lines per month).

3.37 Pilot screen II

Since RNAi mediated knockdown of candidate genes via UAS lines seemed most promising for a large-scale screen after a large library of transgenic flies carrying RNAi silencing constructs under UAS control became publicly available (http://stockcenter.vdrc.at) a second pilot screen was carried together with Malte Kremer to evaluate the potency of RNAi in the experimental system used. Based on the literature (Ethell and Pasquale, 2005; Gao and Bogert, 2003; Guan et al., 2005; Tada and Sheng, 2006) the following lines were selected and screened. A few lines with morphological alterations in the mushroom body calyx or the lobes were identified. *starry night (stan)* and *roundabout (robo)* showed a decrease in microglomerular size and Kenyon cell number appeared to be reduced in *robo*, as suggested by a smaller volume of both calyx and lobes. Upon downregulation of CaMKII (as well as upon overexpression of a dominant negative construct, see above) the shape of the calyx was altered. All Kenyon cells originate from four neuroblasts and it has been shown that the progeny of each of these neuroblasts form clonal units in the calyx, dividing the calyx into four parts. While these subdivisions can easily be seen with genetic labelling they are morphologically not obvious in the wild type. The CaMKII phenotype suggests that the progenies from these four different neuroblasts are not properly integrated into one coherent morphological structure. *stan, pavarotti (pav)* and *sticks and stones (sns)* showed morphological alterations in the mushroom body lobes. The medial lobes were fused in upon downregulation of *stan*. The α'β' lobes were absent and the αβ lobes were reduced in volume in *pav* which might suggest a problem in later divisions of the Kenyon cell neuroblasts. *sns* showed an unusual separation of the tips of the vertical lobes. Microglomerular size was increased upon downregulation of *mef2* or the expression of a dominant negative construct (obtained from Justin Blau, USA). The medial mushroom body lobes were fused. Since the number of animals was low in each case (5-10), the penetrance was not 100% and no quantitative measurements were done these results are suggestive but preliminary.

CG number	Name
1004	rho1, veinlet
1258	pav, pavarotti
1536	sn, singed
2248	Rac1
3064	Futsch
3263	PKA
3936	dNotch
4843	tropomyosin, Tm1
6203	dfmr1
7507	Dhc
8440	Lis1
8556	Rac2, Rac1b, RacB
8556	Rac2
8637	trc, tricornerd
9533	rut, rutabaga
10751	robl, roadblock
11861	dCul3
11865	flamingo, stan, starry night
13521	robo, roundabout
13572	sns, sticks and stones
13752	sns, sticks and stones
14268	dnc, dunce
15112	ena
18069	CaMKII
32045	fry, furry,
74379	PKA

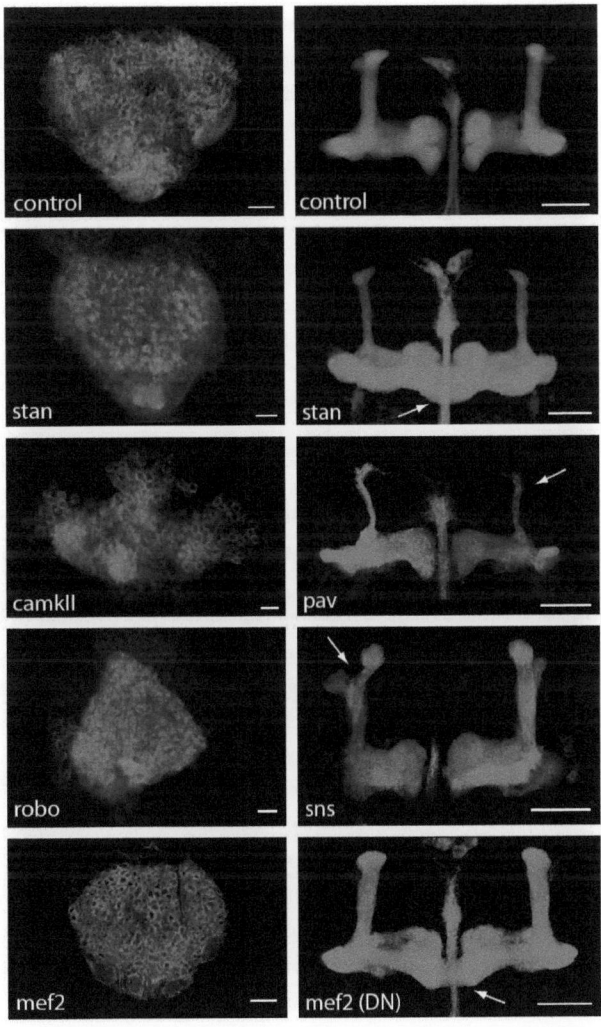

Figure 3.19 | **Mushroom body phenotypes can be induced using RNAi**

Actin-GFP was expressed using the *ok107*-GAL4 driver. Single confocal section of the calyx (left panels) or projection of confocal stacks spanning the mushroom body lobes (right) are shown. See text for descriptions of the phenotypes. Scale bars: 10 μm and 50 μm.

From the results of the RNAi pilot screen I concluded that RNAi mediated knockdown is possible in the mushroom bodies using the *ok107-GAL4* driver as also reported previously (Kobayashi et al., 2006). Unfortunately, I did not find any obvious examples of RNAi induced phenotypes in the LPTCs. Since it appears unlikely that none of the tested candidate genes plays any crucial role in LPTC dendritogenesis this suggests that RNAi is not similarly effective using the *db331-GAL4* driver. Aiming at increasing the efficacy of RNAi downregulation Irina Hein partially rescreened the lines with the following alterations: Flies were raised at 27°C (instead of 25°C) because it is known that the expression levels obtained with the GAL4/UAS system are temperature dependent and male flies were investigated because efficacy of RNAi might be higher than in females (Frank Schnorrer, personal communication). Irina found alterations in LPTC primary dendrite diameter upon downregulation of *futsch*. While this suggests that RNAi mediated gene knockdown can induce morphological alterations in the LPTCs the efficacy appears to be low as *futsch* remains the only example.

3.38 Tools to quantitatively describe alterations in dendritic morphology

Visual investigation of both LPTC and microglomerular morphology was used to screen candidate lines and allowed to detect possibly interesting candidate lines (see part 2.6 for details). To confirm and further characterize possible phenotypes I needed to develop reliable quantitative measurements. A method to manually determine LPTC spine density from 3D reconstructions of VS1 fragments was developed and a manual method to quantify microglomerular inner area was tested (part 3.23). While the manual quantification of spine density was useful for the anatomical description of LPTCs it appeared desirable to develop faster and better-standardized methods for screening purposes. Manual analysis of microglomerular morphology was unsatisfactory for both descriptive and screening purposes. Since reliable quantitative measurements of microglomerular morphology were critical for the attempts to study structural plasticity in the mushroom bodies a description of the software tools allowing automated image analysis are included elsewhere (see 3.24). However, these tools could similarly be used to determine microglomerular morphology upon genetic manipulations. The accuracy, speed and relative ease of quantitative measurements in the mushroom body calyx with automated image analysis represents major strengths of the system for genetic analysis of dendrite morphogenesis.

3.39 Automated or semi-automated analysis of LPTC morphology

Although more efforts were directed towards the development of automated image analysis of microglomerular morphology initial attempts to automatically analyze important parameters of LPTC morphology were carried out. Both automated and semi-automated approaches were tested.

For fully automated image analysis projection images from confocal stacks spanning small fragments of VS1 dendrites were chosen. Different approaches to automatically trace the dendrite, to determine branching orders, to detect spines and quantify spine morphology and density were tested. It was demonstrated that such an approach might be generally feasible. However, significant additional work would be required to allow such an algorithm to deal with variations in morphology and image quality.

Automatic tracing tools are currently being developed by Friedrich Foerstner (Axel Borst lab, Germany) and Hermann Cuntz (Michael Haeusser lab, UK) and might represent a valuable alternative in the near future. A semi-automatic approach based on Amira software developed by Felix Evers (Mike Bate lab, UK) was tested and could yield satisfactory results. However, the amount of manual work required likely precludes any application in a genetic screen (see 4.18).

Taken together, I tested approaches to specifically manipulate target genes in the LPTCs or Kenyon cells and developed tools to quantitatively describe resulting morphological alterations. A critical evaluation of the methodology will be presented after a summary of my findings and a discussion of the anatomical characterization of the LPTCs and the calycal microglomeruli.

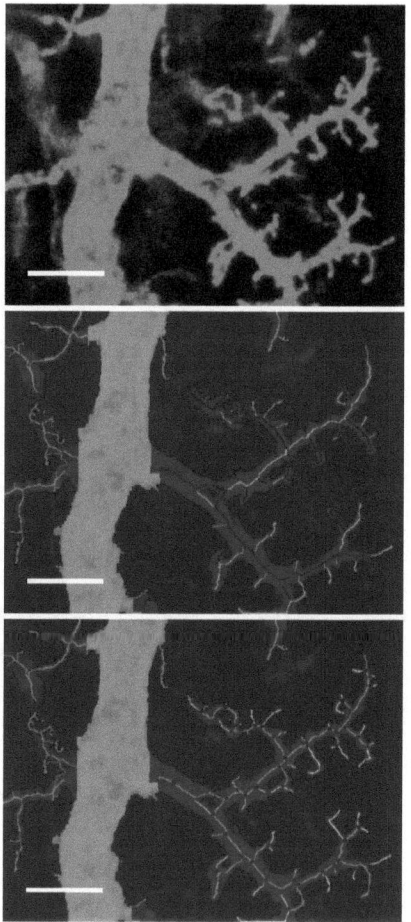

Figure 3.20 | Automated tracing of LPTC dendrites

A projection image through a confocal stack spanning a medial dendritic branch from a VS1 neuron (upper panel) was used to test automated image analysis using Definiens software. A fully automatic algorithm was programmed allowing identification of the primary dendrite (grey), tracing of the dendritic branchlet and assigning branching orders (second, third, fourth and fifth dendritic order are indicated in orange, light blue, dark blue and yellow, respectively (middle panel). Moreover, branching points (red) and terminal protrusions (spines, yellow) were detected. Scale bars: 10 μm.

Discussion

The morphology of dendrites is important to dendritic function and the proper connectivity within neuronal circuits. The factors defining the architecture of dendritic trees include genetic factors as well as neuronal activity. In the recent past, the *Drosophila* PNS has provided valuable insights into the genetic processes underlying dendrite formation. Since the dendrites of the PNS do not receive synaptic input they cannot be used to study essential dendritic properties such as their contribution to synaptogenesis or the relation between morphology and information processing. I was looking for suitable dendrites in the CNS to extend genetic analysis of dendrites in these important regards and to open the possibility to address if experience-dependent structural rearrangements of dendrites do occur in *Drosophila*. In order to identify suitable neurons in the CNS I tested a number of candidates and concluded that Lobula Plate Tangential Cells (LPTCs) and Kenyon cells would best suit the desired purposes.

Together with Ewa Koper, I characterized the morphology of LPTC dendrites in order to investigate if these cells could be used for genetic analysis of dendrite development. Their complex yet stereotyped dendritic trees as well as the availability of specific GAL4 drivers offer great experimental advantages. Moreover, a high abundance of small dendritic protrusions along the higher order branches of LPTC dendrites was noticed. Small dendritic protrusions in flies have been referred to as spines on largely intuitive grounds. It was systematically investigated if *Drosophila* spines share essential characteristics of vertebrate spines. I conclude that LPTC spines morphologically resemble vertebrate spines, are actin-enriched and sites of synaptic contacts. These synaptic contacts are likely excitatory and spine density is altered upon manipulations of the dRac1 levels. I suggest that LPTCs can be used for genetic analysis of dendrite and spine morphogenesis.

Since it appeared unlikely that LPTCs, as part of the circuitry required for wide-field motion detection, might undergo experience-dependent remodelling I also characterized the anatomical properties of Kenyon cell dendrites. Although the poorly-defined morphology of individual Kenyon cells did not encourage studies on structural plasticity I found their dendrites to contribute to microglomerular structures throughout the calyx and the morphology of these microglomeruli was more consistent between animals. Using high-resolution confocal microscopy I analyzed the morphology and connectivity of Kenyon cells, projection neurons and GABAergic interneurons in the mushroom body calyx of *Drosophila* in detail. I showed that each microglomerulus contains a projection neuron bouton and is enveloped by actin-enriched claw-like endings of several Kenyon cells. My data suggest that the Kenyon cell populations contributing to a single microglomerulus can be diverse. Since a better understanding of the synaptic organization in the mushroom body calyx is crucially required to interpret odour representations in this neuropil and consequently also the formation of olfactory associative memories my findings are also of interest to these subjects.

However, the anatomical description of the calyx was aiming at providing the basis for my attempts to address structural plasticity in the fly. It is widely assumed that the formation of long-term memories requires activity-dependent long-lasting morphological alterations in plastic neuronal networks, which might take place in the neuronal spines (Bonhoeffer and Yuste, 2002; Matus, 2005). *Drosophila* mushroom bodies have been implicated in the generation and retrieval of olfactory associative memories (de Belle and Heisenberg, 1994; Heisenberg et al., 1985; Zars et al., 2000). However, the degree to which these plastic processes are associated with structural modifications at the cellular level is not yet resolved. A high enrichment of actin is observed in the microglomeruli (Frambach et al., 2004; Groh et al., 2004) and I could observe a number of small spine-like dendritic protrusions lining the inner rim of the Kenyon cell claw-like ending, thus suggesting these might be potential sites of remodelling. I developed computer-based tools to automatically quantify microglomerular morphology and found indications for rearrangements during the first days after eclosion in the control. I surgically removed third antennal segments and maxillary palps from newly eclosed flies and compared the morphology of their microglomeruli to the wild type. While I cannot exclude very subtle modifications resulting from olfactory deprivation these would be small compared to the developmental rearrangements. Recent improvements of the methodology will allow addressing structural plasticity in more detail in the near future.

Finally, I tried to exploit the anatomical information on LPTC and Kenyon cell dendrites for genetic analysis of dendrites. I identify an experimental setup allowing fast and efficient screening for genes involved in dendrite and spine morphogenesis. I decided to use cell-specific genetic manipulations with the GAL4/UAS system and investigated if the large number of available RNAi lines could be used for genetic manipulations. I carried out two small-scale pilot screens to identify a suitable screening procedure and identified Rac1 as a potential modulator of spine density in the LPTCs and CaMKII and Mef2 as interesting candidate genes involved in dendritogenesis in the calyx. Malte Kremer, Irina Hein and I determined that the potency of RNAi mediated knockdown is likely too low in the LPTCs (using *db331-GAL4*) to reliably induce phenotypes and conclude that a large-scale genetic screen is not feasible using the tested experimental conditions. However, genetic analysis of candidate genes possibly involved in dendritogenesis or structural plasticity in the calyx appears promising.

In the following I will discuss several important aspects of my findings. I will first compare the small dendritic protrusions observed on LPTC dendrites to vertebrate spines. Afterwards, I will comment on the implications of the anatomical description of the mushroom body calyx and about the attempts to use microglomerular complexes in the calyx to study structural plasticity. Finally, I will discuss advantages and disadvantages of using the LPTCs and Kenyon cells for an RNAi based screen aiming at the identification of genetic factors involved in dendritogenesis.

4.1 Spines in *Drosophila*

The presence of processes with spine morphology on dendrites of different types of neurons has been previously described in several types of insects, including cricket and bee (Farris et al., 2001; Hausen et al., 1980; Pierantoni, 1976). *Manduca* motor neurons, for instance, appear to have spine-like protrusions (Weeks and Truman, 1985). Moreover, in cricket's mushroom bodies the presence of synapses on processes that are suggested to resemble spines has been shown by electron microscopy (Frambach et al., 2004). This encouraged Ewa and me to look for spines in *Drosophila*. Compared to other insects as well as to classic systems for spine studies such as vertebrate primary neuronal cultures and slice preparations the merits of *Drosophila* genetics seemed to outweigh the technical limitations due to the small neuronal size. In addition, the existence of tools such as the Gal4/UAS system (Brand and Perrimon, 1993) combined with appropriate reporters allow reaching high levels of imaging resolution with minimal invasiveness.

In *Drosophila*, the presence of spines was suggested by several recent studies, identifying spine-like processes in LPTCs (Reuter et al., 2003; Scott et al., 2002; Scott et al., 2003a) and the presence of synaptic contacts onto small spine-like protrusions in lateral horn neurons (Yasuyama et al., 2003). Nevertheless, none of these studies demonstrated that the observed structures possessed all of the essential characteristics of spines.

A systematic analysis of dendritic spines in LPTCs of *Drosophila* was performed. It was shown that these neurons bear dendritic protrusions that morphologically resemble spines and fall into previously described morphological classes, are actin-enriched and devoid of tubulin and, most importantly, they are sites of synaptic input, which is mostly excitatory. Moreover, they are sensitive to the levels of the actin regulator *dRac1*, suggesting conserved mechanisms of formation and maintenance.

Taken together, these findings led to the conclusion that *Drosophila* has dendritic spines and can thus be used for genetic studies of spines.

4.2 Spine morphology and cytoskeletal organization

In recent years, it has emerged that spines are dynamic processes in culture, in brain slices and *in vivo* (Bonhoeffer and Yuste, 2002; Konur and Yuste, 2004; Oray et al., 2006). Indeed, spines are highly enriched in actin, which is important for their dynamic properties since actin depolymerization blocks spine dynamics (Fischer et al., 1998). Not surprisingly, thus, several of the molecules that are relevant for the regulation of dendrite shape are regulators of actin (Schubert and Dotti, 2007). Profilin, a promoter of actin polymerization, is recruited to spines upon NMDA receptor activation and promotes spine growth (Ackermann and Matus, 2003). Conversely, the severing factor cofilin is inhibited upon LTP induction (Chen et al., 2007) and environment exploration (Fedulov et al., 2007).

Small GTPases of the Rho family have been extensively involved in the regulation of spines through the control of the actin cytoskeleton. In particular, Rac1 is a prominent regulator of spine morphology and density (Luo et al., 1996; Nakayama et al., 2000; Tashiro and Yuste, 2004). Overexpression of a constitutively active version of Rac1 in murine cerebellar Purkinje cells (Luo et al., 1996) and in rat pyramidal neurons (Nakayama et al., 2000) leads to an increase in spine density and a reduction of the spine length (Luo et al., 1996). In contrast, overexpression of dominant negative Rac1 in rat pyramidal neurons (Nakayama et al., 2000) or hippocampal neurons (Zhang and Macara, 2006) results in progressive elimination of dendritic spines.

4.3 Spines in *Drosophila* can be modified genetically

To test the hypothesis that similar genes might be important in controlling spines in both vertebrates and insects the effect of the small GTPase Rac1 on spine density was investigated. In my experiments with *Drosophila* spines the overexpression of wild-type full length dRac1 led to an increase in the number of the spines. Thus, altered levels of Rac1 can modulate the morphology of spines, suggesting that Rac1 might control similar pathways in LPTC spines as in rodent neurons.

While it might appear surprising that overexpression of a dominant negative version of Rac1 leads to the same effect as Rac1 over-expression, this could be explained by the fact that the dominant negative construct can sequester rate-limiting GEFs for other small GTPases, and thus lead to an unspecific effect (Wang and Zheng, 2007). It is worth noting that in *Drosophila* the same axon guidance phenotype was obtained upon overexpression of dominant negative and constitutively active *Rac* (Luo et al., 1994). Initial attempts to carry out RNAi-mediated knock down of Rac1 in standard conditions did not produce alterations (data not shown). Thus, a detailed analysis of the endogenous role of *rac* in LPTC spine regulation awaits more directed experiments.

Taken together, the presented data make the important point that LPTC spines can be modified with genetic tools. Based on the presented characterization, it is now feasible to screen for genetic factors involved in the establishment or in the maintenance of spine morphology, similarly to what has been already done for axon and dendrite morphology in mushroom body neurons (Reuter et al., 2003) and for axonal connectivity in the visual system (Newsome et al., 2000).

4.4 Synaptic contacts onto spines

The major role of spines is to make synaptic contact with their pre-synaptic partners. For this reason it was investigated in detail whether LPTC spines are also sites of synaptic input. In hippocampal CA1 neurons only a fraction of the spines has a mushroom shape which is supposed to represent a mature spine (30%; (Fiala and Harris, 2001). Although it was not possible to make a correlation between maturity of physiological properties and shape of the spine a comparable fraction of mushroom shaped spines on LPTCs (15%), which could also represent mature spines, was observed. By confocal and by serial electron microscopy Ewa Koper observed the presence of synaptic contacts, marked by T-bars, on mushroom shaped spines, but also on spines with different shapes. It is therefore likely that, just as it emerges from vertebrate data (Bourne and Harris, 2008) mushroom-shaped spines are not the only spines that are actively receiving synaptic input.

In agreement with this, by immunohistochemistry I found that almost 90% of the spines examined bear a juxtaposed pre-synaptic marker. In addition, almost all spines contain post-synaptic receptors, as indicated by the localization of the Dα7 subunit of the acetylcholine receptor, revealed both by immunohistochemistry and using a fluorescently-tagged construct. Finally, I have shown that not only pre-synaptic terminals are juxtaposed to actin-rich spines, but I also identified almost 100% juxtaposition of post-synaptic receptors (Dα7) expressed specifically in LPTCs to pre-synaptic labelling. Taken together, these data indicate that a high proportion of spines represent sites of synaptic connectivity.

It is well established that spines are sites of excitatory synaptic input (Gray, 1959). Immunohistochemical and pharmacological experiments performed in *Calliphora* show that LPTC dendrites receive two types of synaptic input: cholinergic excitatory and GABAergic inhibitory (Brotz et al., 1995; Brotz et al., 2001). The Rdl GABA receptor subunit has been previously reported to be localized on the finer dendrite branches of LPTCs (Raghu et al., 2007). It was therefore important to establish the localization of excitatory synapses. The excitatory cholinergic input to LPTCs could be mediated by nicotinic acetylcholine receptors (Brotz et al., 1995). Based on the specific localization of the Dα7 subunit of the acetylcholine receptor, my data strongly suggest that spines receive excitatory synapses. From the non-quantitative electron microscopy analysis it appears that in most cases spines receive a single synaptic input (four cases out of five). Thus, it is conceivable that spines receive mainly excitatory input in these neurons, suggesting that excitatory and inhibitory input could be segregated on the dendrites of LPTCs.

4.5 Functional considerations

What is the function of LPTC spines? Altogether, there is plenty of evidence for plastic processes in the central nervous system of insects (Fahrbach, 2006), such as the morphological changes in the MB of worker honeybees after leaving the hive for their first foraging flight (Durst et al., 1994; Fahrbach et al., 2003; Ismail et al., 2006; Withers et al., 1993) or the volumetric changes in the visual system of *Drosophila* upon different light/dark cycles (Barth et al., 1997). As LPTC spines are highly enriched in actin (Scott et al., 2003b); and the present work) it is possible that LPTC spines are dynamic. Are LPTC spines plastic structures? It has been reported that neither raising flies in the dark, nor eliminating the photoreceptors during development alter the density of these processes on LPTCs (Scott et al., 2003a), suggesting that their formation is independent of sensory input. It is thus possible that the LPTC morphology, which is not affected in these conditions, is under strict genetic regulation. On the other hand, it is conceivable that the direct upstream partner of LPTCs, possibly T4 and T5 neurons (Strausfeld and Lee, 1991), will need to be silenced to completely stop synaptic input to LPTCs (Mizrahi and Libersat, 2002). From own unpublished experiments and from previous work there appear to be many examples of "spiny" neurons in the fly CNS (e.g. (Consoulas et al., 2002; Fayyazuddin et al., 2006). Though no such analysis as the one presented here has been carried out so far it is plausible that dendritic spines are rather widespread in the fly CNS.

Finally, the detailed and quantitative description of the localization of excitatory input onto LPTC dendrites might be interesting for ongoing attempts to understand the information processing underlying motion detection on the computational level (e.g. Borst and Haag, 1996; Farrow et al., 2005; Farrow et al., 2003; Haag and Borst, 2002, 2003; Haag et al., 1997; Haag et al., 1999; Joesch et al., 2008; Raghu et al., 2007).

Based on the data presented here, I consider the LPTCs a very valuable system for detailed analysis of dendritogenesis and spinogenesis. In order to exploit the experimental advantages of LPTCs I tested experimental approaches for efficient genetic analysis of dendrite and spine morphogenesis using cell specific RNAi downregulation of candidate genes. Before discussing these findings, I will turn to the anatomical description of Kenyon cell dendrites and their contributions to microglomerular complexes in the mushroom body calyx and the implications these anatomical findings have. Afterwards I will explain how these findings could be used to study plastic processes at the cellular level before discussing the genetic approaches to dendritogenesis that build on both the anatomical work on the LPTCs and the mushroom body calyx.

4.6 Kenyon cells and calycal microglomeruli

Calycal microglomeruli were identified by electron microscopy in several insect species, including *Drosophila*, and were described to contain a single large presynaptic projection neuron bouton surrounded by numerous small post-synaptic profiles putatively from Kenyon cells and by few GABAergic profiles (Ganeshina and Menzel, 2001; Yasuyama et al., 2002). With the current work I introduce important additions to these early data. I demonstrate first that those small post-synaptic profiles are formed by the claw-like endings of Kenyon cell dendrites and by small spine-like structures protruding from them (Figure 3.9). Second, I show that each claw-like ending enwraps a single projection neuron bouton (Figure 3.10). Third, I identify each microglomerulus as a discrete unit, the boundaries of which are defined by the actin-enriched rim formed by the claw-like endings of several Kenyon cells contacting the projection neuron bouton. The implications of each of these findings are discussed below.

It is important to note that the microglomerular organization of the adult calyx, as described here, appears different than what reported for the larval mushroom body calyx (Masuda-Nakagawa et al., 2005; Ramaekers et al., 2005). In fact, unpublished observations suggest that each of the glomeruli in the larva comprises several microglomeruli and thus that the glomeruli in the larval calyx are a different structure than the microglomeruli in the adult calyx.

4.7 Functional considerations

Electrophysiological recordings in several insect species have shown that the responses of Kenyon cells to odours are sparse, leading to the suggestion that Kenyon cells function as coincidence detectors, responding to coordinate input from projection neurons (Laurent, 2002; Perez-Orive et al., 2002; Szyszka et al., 2005). My morphological and connectivity data are consistent with this possibility.

Around 150-200 projection neurons provide olfactory input to the calyx (Stocker et al., 1990) forming an average of 5 boutons each (Marin et al., 2002). I counted around 1000 microglomeruli in 24 hours old flies. Given technical limitations in my counting methods these counts are approximate. Nevertheless, they are not inconsistent with the previous data, assuming that each microglomerulus contains one, and only one, projection neuron bouton, as our confocal microscopy data indicate. Importantly, they are also supported by 3D reconstructions of calycal microglomeruli obtained from serial-section electron microscopy. This data was obtained by Nancy Butcher and Claudia Groh in Ian Meinertzhagens laboratory as part of a collaboration.

A Kenyon cell has an average of five to seven claw-like endings (Lee et al., 1999; Zhu et al.,

2003; F.L and G.T unpublished). Here, it is demonstrated that each of the claw-like endings contacts a single projection neuron bouton. Although additional input sites cannot be excluded, the data indicate that each Kenyon cell thus receives major input from a very limited number of projection neuron boutons, namely one per claw-like ending. If these boutons originate from different projection neurons, the morphology and connectivity of the Kenyon cells that were described would predict that they could act as detectors of coincident activity in several of their presynaptic partners, and that the number of their presynaptic partners is small compared with the locust (Jortner et al., 2007). Alternatively, all boutons presynaptic to a Kenyon cell could originate from a single projection neuron or from a functionally related set of projection neurons. In that case the functional task of Kenyon cells might be to improve the signal-to-noise ratio in the system.

The synaptic input from projection neurons to any one Kenyon cell is, as suggested from electron microscopy observations, provided at twenty or more active zones. If these sites of synaptic input were distributed evenly among 5-7 claw-shaped endings, each ending would receive three to six sites from any one projection neuron bouton. This number is possibly only what is sufficient to guarantee a reliable postsynaptic response to each incoming presynaptic potential. Based on the anatomical data it can be assumed that each clawed-shaped Kenyon cell ending contacts a different bouton. Thus, an average Kenyon cell could receive input from 5-7 projection neurons, assuming that all of these boutons were from different projection neurons. This estimate matches estimations of the PN:KC convergence ratio based on electrophysiological recordings (Turner et al., 2008). On the other hand, the widespread GABAergic input to a large fraction of microglomeruli revealed by my data (Figure 3.12) and by electron microscopy (Yasuyama et al., 2002) could represent an instrument to keep the single claws silenced until inhibition is relieved. Laurent speculates that a short temporal integration window of Kenyon cells is critical to their specificity and thus to the sparseness of odour representations in the mushroom bodies. He suggests that GABAergic interneurons may provide a periodic reset, preventing temporal integration over successive oscillation cycles of projection neuron activity (Laurent, 2006). I believe that the widespread abundance of GABAergic presynaptic profiles supports these considerations. In honey bee, recurrent GABAergic interneurons innervating the lobes and the calyx (Mobbs, 1982) were suggested to be involved in olfactory memory formation (Grunewald, 1999). It is thus possible that GABAergic neurons in *Drosophila* also have important functions in odour coding and memory formation (Laurent and Naraghi, 1994; Yamazaki et al., 1998). Additional studies will be required to resolve the role of this GABAergic innervation into the calyx and must await more detailed reports on these pathways and further analysis of the physiology of the Kenyon cells and of their dendritic compartments.

4.8 Implications for olfactory representations

Recent evidence indicated that olfactory projection neurons have stereotyped axonal projections in the mushroom body calyx (Jefferis et al., 2007; Lin et al., 2007). Similarly, Kenyon cell dendritic projections appear stereotyped within a number of regions in the calyx (Lin et al., 2007). Thus, a linear processing of olfactory information seemed to be plausible also in the adult mushroom body calyx. These studies, though, rely on registration of projections of neuronal populations in a standard brain model. I concentrated on the microglomerular synaptic complexes and confirm that dendrites of Kenyon cell subpopulations are reliably found at defined positions in the calyx (Strausfeld et al., 2003; Tanaka et al., 2004), where they form distinct populations of microglomeruli (Figure 3.15). Importantly, though, I find that the separation of Kenyon cell subclasses is neither sharp nor complete, consistently with previous results (Tanaka et al., 2004). In fact, different, non-overlapping Kenyon cell subsets contribute to largely overlapping populations of microglomeruli strongly suggesting that an individual projection neuron can contact different classes of Kenyon cells. Since I show that each microglomerulus only contains a single projection neuron bouton this reveals that an individual projection neuron can contact different classes of Kenyon cells. These findings are relevant for an understanding of odour representation because distinct functional roles have been suggested for the different Kenyon cell subsets, such that γ-neurons are involved in short- and $\alpha'\beta'$-neurons in long-term memory (Akalal et al., 2006; Krashes et al., 2007; McGuire et al., 2001; McGuire et al., 2003; Pascual and Preat, 2001; Zars et al., 2000).

4.9 Towards a computational model of the olfactory circuit

Since the olfactory pathway is a relatively simple and well-understood circuit it might be accessible to modelling approaches. It would be highly desirable to simulate the network properties to gain further insights into olfactory coding and the formation of lasting memories. The experimental advantages offered by the genetic accessibility of the pathway could then be used to test predictions obtained from the model (Huerta et al., 2004; Jortner et al., 2007; Laurent, 2002; Nowotny and Huerta, 2003; Nowotny et al., 2005; Smith et al., 2008).

A detailed model of the olfactory pathway would require quantitative anatomical information (Jan Wessnitzer, personal communication). In particular the following questions would be relevant (answers as suggested from the presented work and available literature are indicated in brackets): How many boutons does a projection neuron form on average? (Five to seven.) How many claw-like endings does a Kenyon cell form on average? (Five to seven.) Is the average number of claw-like endings per Kenyon cell similar for all Kenyon cell subpopulations? How many claw-like endings converge per microglomerulus? (Five to fifteen.) Does each microglomerulus contain only a single projection neuron bouton? (Yes.) How many synaptic contacts are there per projection neuron bouton and per Kenyon cell claw-like ending? (Fifty, five to six, respectively.) Are there synaptic contacts between projection neurons and Kenyon cells outside microglomerular complexes? How many GABAergic cells innervate the calyx? Do they represent a homogeneous population or are there functionally distinct subpopulations? Are GABAergic interneurons restricted to the calyx or do they also innervate other brain regions such as the mushroom body lobes or the lateral horn? How many inhibitory neurons contribute to each microglomerulus and how many synapses do they form?

While some of these questions can currently not be answered precisely my work provides an initial framework to conceive suitable experiments to address them more systematically in the future.

4.10 GABAergic interneuron electrophysiology

I demonstrate that processes from *gad1* positive neurons contribute to the vast majority of microglomeruli. My data suggests that these processes are often presynaptic. Immunohistochemistry against the vesicular GABA transporter vGAT suggest that GABAergic presynaptic sites are abundant in the calyx (data not shown, and (Enell et al., 2007; Liu et al., 2007; Yasuyama et al., 2002). Interestingly, I identified a cluster of large *gad1* positive cell bodies in close medial proximity to the calyx that appeared to send processes into the calyx. Colabelling of the mushroom bodies using the GAL4 independent *mb247-DsRed* construct allows the reliable identification of this cluster of neurons based on their genetic labelling (using *gad1-GAL4* and *UAS-mCD8-GFP*) and the proximity to the calyx. The large size of the cells likely allows electrophysiological recordings (Moritz Paehler, personal communication) and backfills could be used to investigate if these neurons indeed innervate the calyx. These findings open the exciting opportunity for an electrophysiological and anatomical characterization of GABAergic interneurons in the calyx. Inhibitory and, more recently, excitatory (Ng et al., 2002; Olsen and Wilson, 2008; Shang et al., 2007; Wilson and Laurent, 2005; Wilson and Mainen, 2006; Wilson et al., 2004) interneurons in the antennal lobe received a lot of attention and have proven to be important for early steps of olfactory information processing. It appears likely that inhibitory interneurons play similarly essential roles for the information processing in the calyx but this possibility has not been addressed to my knowledge.

4.11 Similarities between the mushroom bodies and the mammalian cerebellum

My observations extend previously noted similarities between the circuitry of the mushroom body calyx and the mammalian cerebellum (Schurmann, 1974; Yasuyama et al., 2002). Both the cerebellum and the mushroom body contain large (compared to the entire number of neurons in the mammalian or insect brain) numbers of densely packed neurons, granule cells and Kenyon cells. The axons of both neuronal types run in parallel and form separate structures, the parallel fibres and mushroom body lobes. The dendrites of both neuronal types have an average of five branches and end with characteristic claw-like structures. These claw-like structures are sites of excitatory input from afferent pathways, the mossy fibres or projection neurons. GABAergic interneurons (Golgi cells in the cerebellum) inhibit both granule cells and Kenyon cells. Interestingly, both the cerebellum and the mushroom bodies are important for learning and memory (motor learning and olfactory learning, respectively) and the excitatory input to granule cells (mossy fibres) or Kenyon cells (projection neurons) carries the information about the conditioned stimulus. Both granule cells and Kenyon cells appear to code sparsely and were suggested to function as coincidence detectors (Ito, 2006; Laurent, 2002; Nieus et al., 2006; Ramnani, 2006; Wall, 2005; Xu-Friedman and Regehr, 2003).

In addition, a brain structure in octopus called vertical lobe also shares characteristics with the cerebellum and the mushroom bodies and is also important for learning and memory (Hochner et al., 2006; Shomrat et al., 2008). These similarities raise the possibility that the mammalian cerebellum, the insect mushroom bodies and the vertical lobes in octopus are homologous or analogous structures. It would be very interesting to understand if the computations provided by these similarly organized neuronal networks resemble each other at any level.

The description I provide here will form an important basis for the study of functional properties of the olfactory pathway, including the formation of associative olfactory memory traces. My results provide morphological and connectivity support for physiological data indicating that the response of Kenyon cells to olfactory input is sparse. They furthermore suggest that a proportion of projection neuron boutons contacts more than one type of Kenyon cell, arguing against a linear representation of olfactory information at the level of the mushroom body. The anatomical information is essential for present and future attempts to study structural plasticity in the mushroom body calyx as will be discussed in the following sections.

4.12 Structural plasticity and calycal volume changes

It is well established that actin-based plasticity in dendritic spines, dendritic protein synthesis, synaptic plasticity and long-term memory are intricately correlated in vertebrates (Cingolani and Goda, 2008; Lamprecht and LeDoux, 2004; Losonczy et al., 2008; Matus, 2000; Segal, 2005; Sutton and Schuman, 2006). Since I found Kenyon cell dendrites in the mushroom body calyx to form claw-like endings (Lee et al., 1999; Strausfeld et al., 2003) that bear small dendritic protrusions and are actin-rich, I wondered if these structures could undergo experience-dependent morphological alterations. This hypothesis was encouraged by the role of Kenyon cells in the retrieval of olfactory memories (Gerber et al., 2004; Heisenberg, 2003; Keene and Waddell, 2007) and indications that the volume of the mushroom body calyx is sensitive to the sensory environment in a number of insect species. Calycal volume was shown to increase upon onset of foraging flight in honeybees (Durst et al., 1994; Withers et al., 1993) and it was subsequently demonstrated that this increase in volume has developmental and experience-dependent components (Fahrbach et al., 2003; Ismail et al., 2006) and that it is correlated with alterations in spine density of Kenyon cells (Farris et al., 2001). Age-dependent and task-related changes in calycal volume were also reported in desert ants (Kuhn-Buhlmann and Wehner, 2006). In flies, calycal volume was shown to depend on rearing conditions (Barth and Heisenberg, 1997; Heisenberg et al., 1995). Other indications for structural plasticity in flies are experience-dependent alterations in the antennal lobe (Sachse et al., 2007) – as previously shown in bees (Sigg et al., 1997) – and circadian remodelling of pacemaker neurons (Fernandez et al., 2008) and L2 visual interneurons (Gorska-Andrzejak et al., 2005). Altogether, this suggested that insect brains do exhibit structural remodelling and that the mushroom body calyx in flies might be a promising structure to look for such changes. I considered the actin-rich microglomerular complexes to be possible candidates for remodelling processes in flies.

4.13 Microglomerular rearrangements in other insects

It has been reported that the projection neuron boutons in the calyx of ants increase in volume during aging while their number decreases (Seid et al., 2005). Microglomerular structures in bees were reported to be sensitive to brood temperature in worker bees (Groh et al., 2004) and experience in queens (Groh et al., 2006). Furthermore, the microglomerular size increases upon olfactory/visual associative training in cockroaches (Lent et al., 2007). Seid *et al.* (2005) relate morphological changes in projection neuron bouton size and number with the expansion of the behavioural repertoire during the maturation of a worker ant but these observations remain correlative.

4.14 Evaluation of the methodology

I was aiming at describing microglomerular morphology with a particular emphasis on the postsynaptic side because I hoped to identify dendritic rearrangements upon modifications of the sensory environment. I expressed a GFP-tagged subunit of the Dα7 subunit of the acetylcholine receptor in Kenyon cells to visualize the postsynaptic side of projection neuron to Kenyon cell connections specifically. While this allowed labelling of microglomerular complexes most reliably and consistently, microglomeruli in *Drosophila* remain much less well-defined than in larger insects. Microglomerular morphology was much more comparable between animals than the morphology of single Kenyon cells, but the identification and demarcation of microglomeruli was very difficult and involved subjective judgments.

In order to avoid subjective judgments I developed tools for fully automated analysis of microglomerular morphology. I believe that the impartiality and consistency of the computer algorithm are critical for the intended analysis. Automated image analysis also allowed examination of large data sets which represents another essential prerequisite for reliable investigations on microglomerular morphology.

The most severe problem for algorithm development and evaluation was the absence of a reliable reference point due to the shortcomings of visual analysis.

While my algorithm appears to represent a major progress compared to previous attempts (Krofczik et al., 2008) severe limitations remain. Most notably, the automatic detection is not entirely satisfactory as the rates of both false negatives and false positives are considerable. Moreover, the detection rate depends on the image quality and possibly differs between experimental groups (in particular detection rates seem to be increased at older ages). Finally, all size parameters and numbers were obtained in 2D and systematically deviate from the actual 3D situation.

Due to these limitations I consider most of my results suggestive rather than strictly conclusive. Recent improvements allow circumventing most of the limitations and are currently prepared for publication as a primary research article.

4.15 Improvement of automated image analysis

Future improvements of the Definiens based automatic image analysis could evade some of the current constraints. In particular it would be highly desirable to allow the detection and measurement of microglomeruli in three dimensions. This could likely improve detection accuracy (reduce both false-positives and false-negatives) because it should be more reliable to detect globular structures in 3D instead of ring-like structures in 2D. This would also allow measurements of 3D parameters (bouton volumes instead of areas) and reliable counts. With recent improvements of the Definiens software (XD version) these modifications appear possible but require the development of a new algorithm.

I believe that Definiens software (particularly the new version) allows programming very powerful and sophisticated algorithms. I consider the opportunities offered by automated image analysis to be very valuable (Dragunow, 2008). With the advent of automated microscopes and large scale image acquisition methods (Briggman and Denk, 2006; Huang et al., 2008; Kerr and Denk, 2008; Micheva and Smith, 2007) these approaches will likely gain importance in the future.

4.16 Rearrangements of microglomerular morphology during early adult life

I found indications for alterations in microglomerular morphology during the first days of adult life using both manual measurements and automated image analysis. The area of microglomerular lumens was found to significantly increase from eclosion to day 14 using visual classification and manual tracing. A comparable increase in the ratio of lumen area to total calycal area was obtained with automated image analysis and measurements of brain volume suggested that this change might result from a size increase of microglomeruli (rather than a decrease in total calycal area).

Upon visual investigation, microglomerular morphology appeared to be less ambiguous at later time points facilitating identification and demarcation. This is consistent with higher numbers of microglomeruli detected at later time points with automated image analysis. While it is possible that microglomerular numbers do change over time manual counts suggested a decrease rather than an increase.

Taken together these observations demonstrate alterations of microglomerular morphology during the first days of adult life. These alterations were independently found in control conditions and upon olfactory deprivation using an impartial computer algorithm. Due to limitations of the method it is difficult to unambiguously identify the nature of these morphological rearrangements. My data are most consistent with the assumption that microglomeruli are increasing in size during the first days of adult life while becoming easier to recognize and are possibly more clearly separated. However, I cannot exclude alternative explanations such as changes in microglomerular number.

4.17 Specific manipulations of microglomerular subpopulations

The aim of this set of experiments is to investigate if the morphology of calycal microglomeruli is affected by alterations in neuronal activity. Since I was able to detect morphological rearrangements during aging in the control the method generally appears suitable for this aim. However, major limitations result from the need to normalize for calycal volume to allow compensating for size differences between animals. Recent improvements of the method help circumventing these problems. A primary research article documenting these improvements is currently in preparation.

Building on the morphological description of microglomerular complexes in the mushroom body calyx I demonstrate that the morphology of these structures rearranges during the first days after eclosion. I developed software tools for automated image analysis allowing the unbiased evaluation of large data sets. Combined with genetic labelling allowing the comparison of two distinct populations of microglomeruli it becomes possible to address the effects of neuronal activity on morphology at the cellular level. These tools allow studying possible effects on microglomerular morphology resulting from RNAi mediated downregulation of target genes and could thus help to identify genes involved in dendritogenesis as will be discussed in the following.

4.18 Genetic analysis of dendrite and spine morphogenesis

I was aiming at using the anatomical information on both LPTC and Kenyon cell dendrites for a genetic screen for factors involved in dendritogenesis. I tried to identify an experimental setup combining the virtues of LPTCs and Kenyon cells and allowing fast and efficient screening for genes involved in dendrite and spine morphogenesis. I decided to use cell-specific genetic manipulations with the GAL4/UAS system and investigated if the large number of available RNAi lines could be used for genetic manipulations. I carried out two small-scale pilot screens to identify a suitable screening procedure and identified Rac1 as a potential modulator of spine density in the LPTCs and CaMKII and Mef2 as interesting candidate genes involved in dendritogenesis in the calyx. Malte Kremer, Irina Hein and I determined that the potency of RNAi mediated knockdown is likely too low in the LPTCs (using *db331-GAL4*) to reliably induce phenotypes and conclude that a large-scale genetic screen is not feasible using the tested experimental conditions. However, genetic analysis of candidate genes possibly involved in dendritogenesis or structural plasticity in the calyx appears promising.

4.19 Genetic analysis of dendrite and spine morphogenesis - LPTCs

Looking for suitable neurons to study dendrite morphogenesis in the *Drosophila* CNS I considered the LPTCs excellent candidates for a number of reasons: a) specific GAL4 driver lines allow high-resolution imaging and genetic manipulations, b) LPTCs are individually identifiable and their dendrites are morphologically complex but stereotyped, c) the abundance of published information allows distinguishing reliably between dendritic and axonal compartments and ultimately relating morphological alterations to neuronal function and d) the presence of spine-like structures suggested that factors required for spine morphogenesis could be identified as well as genes involved in dendrite formation.

In order to use LPTC dendrites for genetic analysis I needed methods to efficiently identify candidate genes, to confirm and further characterize interesting phenotypes and to describe them quantitatively. The challenges of each of these steps will be discussed consecutively.

4.20 An efficient screening procedure to identify candidate genes

The main advantage of studying dendrite and spine morphogenesis in *Drosophila* is the possibility to identify genes involved in these processes at a much faster pace than in most other systems. Large and unbiased genetic screens as well as systematic tests of long lists of candidate genes appear feasible in flies and this possibility represents the major motivation in developing a dendrite model in the fly CNS. However, assessing the morphology of complex dendritic trees requires high-resolution imaging at the confocal microscope and the amount of work for sample preparation and investigation is considerable. Two pilot screens to establish and test a screening protocol were carried out (the second together with Malte Kremer). I estimate that 100 candidate lines could be analyzed per month if 5-10 brains per line are prepared and visually investigated for a couple of seconds to minutes each. LPTC dendrites appear to be sufficiently stereotyped to allow indentifying genes affecting dendritic branching or guidance as well as spine morphology or density – unless the effect is rather subtle or penetrance is low. Subtle phenotypes will be in the range of normal inter-animal variation and thus be very difficult to detect. I identified Rac1 as a possible modulator of spine density illustrating that it is possible to find interesting genes using fast visual investigation under screening conditions. However, I believe that only phenotypes similarly strong as (or stronger than) the Rac1 phenotype can be identified and consider the amount of false negatives to be likely considerable. While this clearly represents a major constraint, more careful investigation of possible phenotypes would require the acquisition and quantitative evaluation of representative images. The former is time-consuming and the latter technically very challenging on a large scale (see below).

4.21 Cell-specific genetic manipulations are required for efficient genetic analysis

A variety of means to induce mutations or to manipulate genes has been used in *Drosophila* (St Johnston, 2002). I decided to use cell specific genetic manipulations of target genes based on the GAL4/UAS system (Brand and Perrimon, 1993) for two reasons. First, the efficacy of genetic manipulations appeared higher than with other means and second, the restriction of the manipulation of target genes to few cells allowed circumventing lethality due to essential functions of candidate genes in other tissues and increased the probability to identify genes with cell-autonomous functions.

Alternative approaches such as random mutagenesis with EMS or transposon-mediated gene disruption (Spradling et al., 1999) would likely yield interesting mutations at too low rates given the laborious screening procedure and the probably high rate of false negatives. A number of constitutive active or dominant negative variants of interesting candidate genes were available under UAS control and tested in a first pilot screen. I concluded that it is possible to genetically manipulate LPTCs using suitable UAS constructs and to identify resulting phenotypes. For genetic analysis on a larger scale the recently established library of UAS based RNAi constructs appeared most promising because it allows the individual and cell-specific downregulation of around 80% genes of the *Drosophila* genome (Dietzl et al., 2007).

4.22 Low efficacy of RNAi mediated gene knockdown in the LPTCs

Malte Kremer and I tested the potency of RNAi mediated gene knockdown in the LPTCs using the *db331-GAL4* driver and a list of around 30 candidate RNAi lines selected based on published functions in dendrite formation. Most candidate lines were tested both with and without coexpression of Dicer2 which is commonly used to increase the efficacy of RNAi knockdown. No obvious phenotype was detected. Irina Hein partially rescreened the lines at higher temperature (27 °C instead of 25 °C) and looked at males instead of females in order to improve the efficacy of RNAi knockdown (Frank Schnorrer, Germany, personal communication). Irina identified a subtle phenotype of primary dendrite diameter upon *futsch* downregulation. Since this was the only effect obtained with an RNAi line and since it appeared unlikely that no other gene of the candidate set is involved in LPTC dendrite formation I concluded that the potency of RNAi mediated gene downregulation is likely too low in LPTCs to reliably induce phenotypes. This conclusion may be supported by the failure to downregulate the expression of mCD8GFP via coexpression of two different RNAi lines targeting GFP. My observations with *rac1* could not be reproduced using RNAi. A suitable line to confirm the published effect of *cdc42* on LPTC dendrites was unfortunately not available.

The RNAi library has been used by several labs and generally appears to work reliably and for the majority of genes (Dietzl et al., 2007). The low efficacy in the LPTCs might result from fairly late onset of strong *db331-GAL4* expression. Although *db331-GAL4* can be used to visualize LPTCs throughout the second half of pupal development (Ewa Koper, personal communication) expression levels are weak and only increase shortly prior to eclosion. Dendrites and spines have almost reached their adult morphology at that time and might no longer be susceptible to genetic manipulation.

4.23 Additional problems for large scale genetic analysis of LPTC dendrites

The low efficacy of RNAi in the LPTCs suggests that large-scale genetic analysis of dendritogenesis is impossible using the tested experimental conditions. Moreover, two additional problems pose further constraints on genetic analysis: difficulties in reliable and fast quantifications of possible phenotypes and difficulties in confirming effects obtained with RNAi via mutant analysis.

Methods for manual quantification of spine density and length were but more standardized and reliable quantification approaches would be desirable for the analysis of many candidate lines. These would possibly also allow reducing the number of false negatives during screening. Automated approaches to quantify spine density and shape have been used in mammals and appear encouraging although not directly applicable (Cheng et al., 2007; Shen et al., 2008). Initial attempts to develop automated image analysis approaches based on Definiens software appeared promising but revealed significant challenges and were not carried any further. Alternative approaches for automatic tracing of dendritic branches are currently developed by Friedrich Foerster (Axel Borst lab, Germany) and Hermann Cuntz (Michael Haeusser lab, UK) and might represent a valuable tool in the near future. Semi-automatic tracing tools have been developed by Felix Evers (Mike Bate lab, UK)(Evers et al., 2006; Evers et al., 2005; Schmitt et al., 2004) and were tested but were time-consuming and not suitable for application on a large scale. In summary, the difficulties of reliable and fast quantitative measurements of dendrite and spine morphology represent substantial challenges for a large scale screen.

Successfully identified candidate genes would have to be characterized further to confirm the phenotype and describe it in more detail. This requires the analysis of mutants. It is possible to apply MARCM to study the cell-autonomous function of genes in the LPTCs specifically. However, the frequency of MARCM clones appears to be very low (Ewa Koper, Irina Hein,

Shamprasad Raghu, personal communication) and this precludes fast and efficient characterization of candidate genes.

4.24 Genetic analysis of dendrite morphogenesis – calycal microglomeruli

Individual Kenyon cell dendrites appear not well suited for the identification of genes involved in dendritogenesis but the morphology of microglomerular complexes is roughly stereotyped between animals. Since microglomeruli contain actin-rich claw-like endings from several Kenyon cell dendrites and their small dendritic protrusions I considered their morphology to be a potentially interesting indicator for structural alterations in large populations of Kenyon cells. Since it was possible to combine the *db331-GAL4* (LPTCs) and *ok107-GAL4* (mushroom bodies) drivers I used both systems simultaneously. Visual investigation of the mushroom bodies required only a couple of seconds per brain and no additional time for sample preparation was needed. The experimental setup was thus identical with the setup for LPTC analysis as described above.

While I only identified one interesting candidate gene, *rac1*, to affect the LPTCs, the number of genes affecting mushroom body morphology was generally higher. Importantly, it was possible to induce interesting phenotypes with RNAi using the *ok107-GAL4* driver. I conclude that it is possible to induce and identify possibly interesting phenotypes with suitable UAS constructs – including RNAi. The strongest and most interesting phenotypes were observed with CaMKII and Mef2. Manipulation of CaMKII using either RNAi or a dominant negative construct (obtained from Leslie Griffith, USA) disturbed the morphological integrity of the calyx. While the calyx is usually a single, compact structure I observed the segregation into four substructures upon manipulations of CaMKII function. Since Kenyon cells are known to derive from four neuroblasts this might suggest that they fail to integrate into a unified structure. The morphology of microglomerular complexes was not obviously affected.

The strongest effects on microglomerular morphology were observed upon manipulations of *mef2* using either RNAi or a dominant negative construct (obtained from Justin Blau, USA). Microglomerular size appeared to be increased and the morphology of the calyx was occasionally affected. The medial mushroom body lobes were fused. The phenotypes might be interesting because of the implication of CaMKII and Mef2 in synaptic plasticity and structural plasticity, respectively (see below). Specific molecular tools including dominant negative and overexpression constructs (Leslie Griffith, USA and Mike Taylor, UK) are available and could facilitate future analysis. YFP-tagged CaMKII variants (Sam Kunes, USA) have been used in the antennal lobe and appear to localize to microglomeruli in the calyx (data not shown).

4.25 Further characterization of mushroom body phenotypes

Since automated image analysis tools have been developed for quantitative description of calycal morphology it should be relatively easy to further characterize candidate genes affecting calycal morphology – such as *camkII* and *mef2*. The possibility to consistently analyze large data sets with an unbiased and fast computer algorithm represents a major advantage for genetic analysis of dendrites in the calyx. The constraints imposed by the necessity for time-consuming manual analysis of both LPTC and PNS dendrites could potentially be circumvented.

Moreover, mutant analysis in the calyx is much easier than in other systems because the frequency of MARCM clones is comparatively high. Since many of the interesting candidate genes (*fmr1, appl, rut, dnc*, see below) are viable, an initial characterization would not even require MARCM experiments.

4.26 Potentially interesting candidate genes

Taken together, it appears promising to investigate calycal morphology upon manipulations of a number of candidate genes in more detail. *camkII, mef2* and some prominent learning mutants (*dnc, rut, amn*) are considered most promising.

Mef2

In mammals, the MEF2 family of transcription factors is highly expressed in the brain when neurons undergo dendritic maturation and synapse formation (Lyons et al., 1995). MEF2A is especially abundant in granule neurons of the cerebellar cortex throughout the period of synaptogenesis (Lyons et al., 1995) and it has been shown to play a key role in the morphogenesis of granule neuron dendritic claws (Shalizi et al., 2006). This is particularly interesting because of numerous similarities between granule neurons in the mammalian cerebellar cortex and Kenyon cells in insects (see part 4.11). Moreover, it has been shown that mammalian MEF2 transcription factors negatively regulate excitatory synapse number in an activity-dependent manner during synaptic development *in vitro* (Flavell et al., 2006). In *Drosophila* (as in vertebrates), Mef2 has been implicated in muscle development. It is likely specifically enriched in the mushroom bodies since the (mushroom body specific) *mb247* enhancer trap is inserted into the Mef2 locus (Hiromu Tanimoto, personal communication).

CaMKII

Ca^{2+}/calmodulin-dependent protein kinase II (CaMKII), a multifunctional serine/threonine kinase, is enriched at synapses, especially at the postsynaptic density. In mice, CaMKII has been shown to be central to the regulation of glutamatergic synapses. Several lines of evidence indicate that CaMKII detects Ca^{2+} elevation and initiates the biochemical cascade that potentiates synaptic transmission during LTP. The enzyme might be sufficient for persistence of LTP. Mutant mice show strong memory impairments in behavioural tests. Moreover, CaMKII appears to regulate the formation of synapses and dendritic spines (Colbran and Brown, 2004; Lisman et al., 2002).

CaMKII homologues have been found in all multicellular organisms that have been examined. In *Drosophila*, it has been shown that CaMKII is required for both behavioural plasticity in courtship learning (Mehren and Griffith, 2006) and for normal synaptic function at the neuromuscular junction (Griffith and Budnik, 2006). It also alters structural plasticity and cytoskeletal dynamics in flies (Andersen et al., 2005).

Given its very prominent role in molecular studies of learning and memory in mammals, CaMKII appears to be a very interesting candidate for detailed analysis in the mushroom body calyx because it might also be involved in structural plasticity.

dFmr1

Fragile X syndrome is a common form of inherited mental retardation that occurs with a frequency of 1 to 4000 in males and 1 to 8000 in females. Pathology in the brain of both human patients and *fmr1* knockout mice appears to be limited to the presence of abnormal dendritic spines reminiscent of a delay in their maturation (Kaytor and Orr, 2001) In *Drosophila* the single homologue of *fmr1* was implicated in regulation of synapse growth and function on both sides of the synaptic cleft, primarily through its regulation of the expression of the microtubule-associated protein Futsch, likely as a translational repressor (Kaytor and Orr, 2001; Zhang et al., 2001).

Pan *et al.* (Pan et al., 2004) demonstrate that in Kenyon Cells, FMRP (fragile X mental retardation protein) bidirectionally regulates multiple levels of structural architecture, including process formation from the soma, dendritic elaboration, axonal branching, and synaptogenesis. Charles Tessier and Kendal Broadie (Tessier and Broadie, 2008) demonstrated that FMRP expression coincides with a transient window of late brain development. During this time, FMRP is positively regulated by sensory input activity and is required to limit axon growth and for efficient activity-dependent pruning of axon branches in the mushroom body. The authors suggest that FMRP has a primary role in activity-dependent neural circuit refinement during late brain development (Charles Tessier, personal communication).

Taken together these findings make *fmr1* a very interesting candidate to test for morphological alterations in Kenyon Cell dendrites throughout development.

dunce, rutabaga, amnesiac

Mutants defective in olfactory associative learning appear to be good candidates to be tested for alterations in microglomerular morphology and plasticity. The first two genes identified from unbiased screens for learning and memory defects uncovered two components of the cyclic AMP (cAMP) cascade — *dunce (dnc)*, encoding cAMP phosphodiesterase, and *rutabaga (rut)*, encoding a type I Ca2+/calmodulin-stimulated adenylyl cyclase and both proteins were shown to be selectively enriched in the mushroom bodies. Rutabaga has received the most attention in memory studies because it could plausibly register coincident activity following both conditioned and unconditioned stimuli in specific Kenyon Cells through Ca^{2+} influx following projection neuron-driven depolarization and monoamine binding to G-protein coupled receptors. Selective expression of a *rut* cDNA in the mushroom bodies, even only in adulthood, restores wild-type memory to rut mutant flies (McGuire et al., 2003; Zars et al., 2000), suggesting that *rut*-dependent plasticity in the mushroom bodies is sufficient for olfactory memory (Keene and Waddell, 2007).

Another well studied learning mutant is *amnesiac (amn)* which encodes for a putative neuropeptide expressed in DPM (dorsal paired medial) neurons. Expressing Amn in DPM neurons of a mutant *amn* fly restores aversive and appetitive olfactory memory, suggesting that DPM neurons are a crucial site of amn action in the brain to stabilize both forms of olfactory memory (Keene et al., 2004; Waddell et al., 2000).

Barth and Heisenberg (Barth and Heisenberg, 1997) found calycal volume to be affected by visual experience and demonstrate that this effect depends on both *dnc* and *amn*, but not on *rut*. All three mutants were implicated in structural plasticity observed as a consequence of different population densities (Barth and Heisenberg, 1997; Hitier et al., 1998).

Taken together the intended genetic analysis of dendrite and spine morphogenesis in the LPTCs appears to be more difficult than anticipated. Onset of strong *db331-GAL4* expression might be too late to reliably induce RNAi mediated gene knockdown. Moreover, the evaluation of potential phenotypes relies on tedious and time-consuming manual analysis and is more difficult than similar approaches used for PNS dendrites. Finally, the low frequency of MARCM clones in the LPTCs precludes the efficient characterization of candidate genes in more detail. Using the tested experimental conditions a large-scale screen for factors involved in dendrite and spine formation is not feasible. However, the experimental advantages of the systems could be used for detailed investigations of dendritic development including spine and synapse formation.

Genetic analysis of dendritogenesis in the calyx appears more promising. Genes involved in dendrite formation, in defining microglomerular morphology and structural plasticity could be identified. Interesting phenotypes were obtained with RNAi suggesting that the large RNAi library could be used. Phenotypes were identified after brief visual investigation. Recently developed computer-based quantification could replace visual investigation or assist more detailed morphological analysis of selected candidates. Several interesting candidate genes were identified and *camkII* and *mef2* appear particularly promising. Based on the literature, additional potentially interesting candidates are *fmr1, dnc, rut* and *amn*.

4.27 Conclusion

The presented data opens a wide variety of possibilities to study dendrites in the *Drosophila* CNS. Lobula Plate Tangential Cells appear particularly suitable to study developmental aspects of dendritogenesis. Importantly, the identifications of small protrusions along their dendrites that share essential characteristics with vertebrate spines open the opportunity to study spine morphogenesis in the fly. Kenyon cell dendrites are of interest because they contribute to microglomerular structures present throughout the mushroom body calyx. Since the morphology of microglomeruli is roughly consistent between animals and since the volume of the mushroom bodies has been shown to depend on experience in a number of insect species including the fly it appears plausible that microglomeruli might undergo experience-dependent alterations. Although evidence supporting this possibility has as yet not been obtained the development of new genetic tools and computer-based quantification methods provide an essential basis for ongoing more detailed investigations.

In order to identify genes involved in dendrite and spine morphogenesis experimental approaches suitable for a large-scale genetic screen were tested. While an RNAi based downregulation of target genes appears to be difficult with the *db331-GAL4* driver in the LPTCs this approach appears very promising in the Kenyon cells. Promising candidates affecting Kenyon cell morphology have been identified and raise the prospect of studying the genetic basis of dendrite development.

Besides these opportunities to use the morphological information on LPTCs and Kenyon cells to study different aspects of dendrites in the *Drosophila* CNS the morphology of these neurons is also critically relevant to their unique functions. The anatomical findings on the LPTCs and Kenyon cells are thus also relevant for attempts to describe the functions of wide-field motion detection and of the olfactory circuit, respectively.

Literature

Since I did not want to omit references that might be of use for the expert reader the following list of references fails to be limited to the essential. The use of gray font for dispensable information will hopefully help identifying the most valuable references.

5.1 Books

Dendrites
Stuart G, Spruston N, Haeusser M (editors)
Oxford University Press, 2008

Invertebrate Neurobiology
North G, Greenspan RJ (editors)
CSHL Press, 2007

An Introduction to Nervous Systems
Greenspan, RJ
CSHL Press, 2006

Brain Development in Drosophila melanogaster
Technau GM (editor)
Springer, 2008

Fly Pushing
Greenspan, RJ
CSHL Press, 2004

Memory
Squire, R and Kandel, ER
Roberts and Company, 2009

5.2 Articles

Ackermann, M., and Matus, A. (2003). Activity-induced targeting of profilin and stabilization of dendritic spine morphology. Nat Neurosci 6, 1194-1200.

Akalal, D.B., Wilson, C.F., Zong, L., Tanaka, N.K., Ito, K., and Davis, R.L. (2006). Roles for Drosophila mushroom body neurons in olfactory learning and memory. Learn Mem 13, 659-668.

Allen, M.J., Drummond, J.A., and Moffat, K.G. (1998). Development of the giant fiber neuron of Drosophila melanogaster. J Comp Neurol 397, 519-531.

Allen, M.J., Godenschwege, T.A., Tanouye, M.A., and Phelan, P. (2006). Making an escape: development and function of the Drosophila giant fibre system. Semin Cell Dev Biol 17, 31-41.

Alvarez, V.A., and Sabatini, B.L. (2007). Anatomical and physiological plasticity of dendritic spines. Annu Rev Neurosci 30, 79-97.

Andersen, R., Li, Y., Resseguie, M., and Brenman, J.E. (2005). Calcium/calmodulin-dependent protein kinase II alters structural plasticity and cytoskeletal dynamics in Drosophila. J Neurosci 25, 8878-8888.

Ashraf, S.I., McLoon, A.L., Sclarsic, S.M., and Kunes, S. (2006). Synaptic protein synthesis associated with memory is regulated by the RISC pathway in Drosophila. Cell 124, 191-205.

Assisi, C., Stopfer, M., Laurent, G., and Bazhenov, M. (2007). Adaptive regulation of sparseness by feedforward inhibition. Nat Neurosci 10, 1176-1184.

Axel, R. (1995). The molecular logic of smell. Sci Am 273, 154-159.

Baas, P.W., and Buster, D.W. (2004). Slow axonal transport and the genesis of neuronal morphology. J Neurobiol 58, 3-17.

Bailey, C.H., Chen, M., Keller, F., and Kandel, E.R. (1992). Serotonin-mediated endocytosis of apCAM: an early step of learning-related synaptic growth in Aplysia. Science 256, 645-649.

Baines, R.A., Robinson, S.G., Fujioka, M., Jaynes, J.B., and Bate, M. (1999). Postsynaptic expression of tetanus toxin light chain blocks synaptogenesis in Drosophila. Curr Biol 9, 1267-1270.

Barth, M., and Heisenberg, M. (1997). Vision affects mushroom bodies and central complex in Drosophila melanogaster. Learn Mem 4, 219-229.

Barth, M., Hirsch, H.V., Meinertzhagen, I.A., and Heisenberg, M. (1997). Experience-dependent developmental plasticity in the optic lobe of Drosophila melanogaster. J Neurosci 17, 1493-1504.

Berdnik, D., Chihara, T., Couto, A., and Luo, L. (2006). Wiring stability of the adult Drosophila olfactory circuit after lesion. J Neurosci 26, 3367-3376.

Bliss, T.V., and Lomo, T. (1973). Long-lasting potentiation of synaptic transmission in the dentate area of the anaesthetized rabbit following stimulation of the perforant path. J Physiol *232*, 331-356.

Bloodgood, B.L., and Sabatini, B.L. (2007). Ca(2+) signaling in dendritic spines. Curr Opin Neurobiol *17*, 345-351.

Bonhoeffer, T., and Yuste, R. (2002). Spine motility. Phenomenology, mechanisms, and function. Neuron *35*, 1019-1027.

Borst, A., and Egelhaaf, M. (1992). In vivo imaging of calcium accumulation in fly interneurons as elicited by visual motion stimulation. Proc Natl Acad Sci U S A *89*, 4139-4143.

Borst, A., and Haag, J. (1996). The intrinsic electrophysiological characteristics of fly lobula plate tangential cells: I. Passive membrane properties. J Comput Neurosci *3*, 313-336.

Borst, A., and Haag, J. (2002). Neural networks in the cockpit of the fly. J Comp Physiol A Neuroethol Sens Neural Behav Physiol *188*, 419-437.

Bourne, J.N., and Harris, K.M. (2008). Balancing Structure and Function at Hippocampal Dendritic Spines. Annu Rev Neurosci.

Boyle, M., Nighorn, A., and Thomas, J.B. (2006). Drosophila Eph receptor guides specific axon branches of mushroom body neurons. Development (Cambridge, England) *133*, 1845-1854.

Brand, A.H., and Perrimon, N. (1993). Targeted gene expression as a means of altering cell fates and generating dominant phenotypes. Development *118*, 401-415.

Brenman, J.E., Gao, F.B., Jan, L.Y., and Jan, Y.N. (2001). Sequoia, a tramtrack-related zinc finger protein, functions as a pan-neural regulator for dendrite and axon morphogenesis in Drosophila. Developmental cell *1*, 667-677.

Briggman, K.L., and Denk, W. (2006). Towards neural circuit reconstruction with volume electron microscopy techniques. Curr Opin Neurobiol *16*, 562-570.

Broome, B.M., Jayaraman, V., and Laurent, G. (2006). Encoding and decoding of overlapping odor sequences. Neuron *51*, 467-482.

Brotz, T.M., Egelhaaf, M., and Borst, A. (1995). A preparation of the blowfly (Calliphora erythrocephala) brain for in vitro electrophysiological and pharmacological studies. J Neurosci Methods *57*, 37-46.

Brotz, T.M., Gundelfinger, E.D., and Borst, A. (2001). Cholinergic and GABAergic pathways in fly motion vision. BMC Neurosci *2*, 1.

Broughton, S.J., Kitamoto, T., and Greenspan, R.J. (2004). Excitatory and inhibitory switches for courtship in the brain of Drosophila melanogaster. Curr Biol *14*, 538-547.

Brown, S.L., Joseph, J., and Stopfer, M. (2005). Encoding a temporally structured stimulus with a temporally structured neural representation. Nat Neurosci *8*, 1568-1576.

Buck, L.B. (2000). The molecular architecture of odor and pheromone sensing in mammals. Cell *100*, 611-618.

Campbell, R.E., Tour, O., Palmer, A.E., Steinbach, P.A., Baird, G.S., Zacharias, D.A., and Tsien, R.Y. (2002). A monomeric red fluorescent protein. Proc Natl Acad Sci U S A *99*, 7877-7882.

Chen, L.Y., Rex, C.S., Casale, M.S., Gall, C.M., and Lynch, G. (2007). Changes in synaptic morphology accompany actin signaling during LTP. J Neurosci *27*, 5363-5372.

Chen, Y., and Ghosh, A. (2005). Regulation of dendritic development by neuronal activity. J Neurobiol *64*, 4-10.

Cheng, J., Zhou, X., Miller, E., Witt, R.M., Zhu, J., Sabatini, B.L., and Wong, S.T. (2007). A novel computational approach for automatic dendrite spines detection in two-photon laser scan microscopy. Journal of neuroscience methods *165*, 122-134.

Cingolani, L.A., and Goda, Y. (2008). Actin in action: the interplay between the actin cytoskeleton and synaptic efficacy. Nat Rev Neurosci *9*, 344-356.

Cline, H.T. (2001). Dendritic arbor development and synaptogenesis. Curr Opin Neurobiol *11*, 118-126.

Colbran, R.J., and Brown, A.M. (2004). Calcium/calmodulin-dependent protein kinase II and synaptic plasticity. Current opinion in neurobiology *14*, 318-327.

Connolly, J.B., Roberts, I.J., Armstrong, J.D., Kaiser, K., Forte, M., Tully, T., and O'Kane, C.J. (1996). Associative learning disrupted by impaired Gs signaling in Drosophila mushroom bodies. Science *274*, 2104-2107.

Consoulas, C., Restifo, L.L., and Levine, R.B. (2002). Dendritic remodeling and growth of motoneurons during metamorphosis of Drosophila melanogaster. J Neurosci *22*, 4906-4917.

Couto, A., Alenius, M., and Dickson, B.J. (2005). Molecular, anatomical, and functional organization of the Drosophila olfactory system. Curr Biol *15*, 1535-1547.

Crittenden, J.R., Skoulakis, E.M., Han, K.A., Kalderon, D., and Davis, R.L. (1998). Tripartite mushroom body architecture revealed by antigenic markers. Learn Mem *5*, 38-51.

Datta, S.R., Vasconcelos, M.L., Ruta, V., Luo, S., Wong, A., Demir, E., Flores, J., Balonze, K., Dickson, B.J., and Axel, R. (2008). The Drosophila pheromone cVA activates a sexually dimorphic neural circuit. Nature *452*, 473-477.

de Belle, J.S., and Heisenberg, M. (1994). Associative odor learning in Drosophila abolished by chemical ablation of mushroom bodies. Science *263*, 692-695.

de Bruyne, M., Foster, K., and Carlson, J.R. (2001). Odor coding in the Drosophila antenna. Neuron *30*, 537-552.

Dent, E.W., and Gertler, F.B. (2003). Cytoskeletal dynamics and transport in growth cone motility and axon guidance. Neuron *40*, 209-227.

Dickson, B.J. (2002). Molecular mechanisms of axon guidance. Science *298*, 1959-1964.

Dietzl, G., Chen, D., Schnorrer, F., Su, K.C., Barinova, Y., Fellner, M., Gasser, B., Kinsey, K., Oppel, S., Scheiblauer, S., et al. (2007). A genome-wide transgenic RNAi library for conditional gene inactivation in Drosophila. Nature *448*, 151-156.

Dimitrova, S., Reissaus, A., and Tavosanis, G. (2008). Slit and Robo regulate dendrite branching and elongation of space-filling neurons in Drosophila. Dev Biol *324*, 18-30.

Dragunow, M. (2008). High-content analysis in neuroscience. Nat Rev Neurosci *9*, 779-788.

Dudai, Y. (1996). Consolidation: fragility on the road to the engram. Neuron *17*, 367-370.

Dunaevsky, A., Tashiro, A., Majewska, A., Mason, C., and Yuste, R. (1999). Developmental regulation of spine motility in the mammalian central nervous system. Proc Natl Acad Sci U S A *96*, 13438-13443.

Durst, C., Eichmuller, S., and Menzel, R. (1994). Development and experience lead to increased volume of subcompartments of the honeybee mushroom body. Behav Neural Biol *62*, 259-263.

Dutta, D., Bloor, J.W., Ruiz-Gomez, M., VijayRaghavan, K., and Kiehart, D.P. (2002). Real-time imaging of morphogenetic movements in Drosophila using Gal4-UAS-driven expression of GFP fused to the actin-binding domain of moesin. Genesis *34*, 146-151.

Edwards, K.A., Demsky, M., Montague, R.A., Weymouth, N., and Kiehart, D.P. (1997). GFP-moesin illuminates actin cytoskeleton dynamics in living tissue and demonstrates cell shape changes during morphogenesis in Drosophila. Dev Biol *191*, 103-117.

Enell, L., Hamasaka, Y., Kolodziejczyk, A., and Nassel, D.R. (2007). gamma-Aminobutyric acid (GABA) signaling components in Drosophila: immunocytochemical localization of GABA(B) receptors in relation to the GABA(A) receptor subunit RDL and a vesicular GABA transporter. J Comp Neurol *505*, 18-31.

Engert, F., and Bonhoeffer, T. (1999). Dendritic spine changes associated with hippocampal long-term synaptic plasticity. Nature *399*, 66-70.

Ethell, I.M., and Pasquale, E.B. (2005). Molecular mechanisms of dendritic spine development and remodeling. Prog Neurobiol *75*, 161-205.

Evers, J.F., Muench, D., and Duch, C. (2006). Developmental relocation of presynaptic terminals along distinct types of dendritic filopodia. Dev Biol *297*, 214-227.

Evers, J.F., Schmitt, S., Sibila, M., and Duch, C. (2005). Progress in functional neuroanatomy: precise automatic geometric reconstruction of neuronal morphology from confocal image stacks. J Neurophysiol *93*, 2331-2342.

Fahrbach, S.E. (2006). Structure of the mushroom bodies of the insect brain. Annu Rev Entomol *51*, 209-232.

Fahrbach, S.E., Farris, S.M., Sullivan, J.P., and Robinson, G.E. (2003). Limits on volume changes in the mushroom bodies of the honey bee brain. J Neurobiol *57*, 141-151.

Farris, S.M. (2005). Developmental organization of the mushroom bodies of Thermobia domestica (Zygentoma, Lepismatidae): insights into mushroom body evolution from a basal insect. Evolution & development 7, 150-159.

Farris, S.M., Abrams, A.I., and Strausfeld, N.J. (2004). Development and morphology of class II Kenyon cells in the mushroom bodies of the honey bee, Apis mellifera. J Comp Neurol 474, 325-339.

Farris, S.M., Robinson, G.E., and Fahrbach, S.E. (2001). Experience- and age-related outgrowth of intrinsic neurons in the mushroom bodies of the adult worker honeybee. J Neurosci 21, 6395-6404.

Farris, S.M., and Strausfeld, N.J. (2003). A unique mushroom body substructure common to basal cockroaches and to termites. J Comp Neurol 456, 305-320.

Farrow, K., Borst, A., and Haag, J. (2005). Sharing receptive fields with your neighbors: tuning the vertical system cells to wide field motion. J Neurosci 25, 3985-3993.

Farrow, K., Haag, J., and Borst, A. (2003). Input organization of multifunctional motion-sensitive neurons in the blowfly. J Neurosci 23, 9805-9811.

Fayyazuddin, A., Zaheer, M.A., Hiesinger, P.R., and Bellen, H.J. (2006). The nicotinic acetylcholine receptor Dalpha7 is required for an escape behavior in Drosophila. PLoS Biol 4, e63.

Fedulov, V., Rex, C.S., Simmons, D.A., Palmer, L., Gall, C.M., and Lynch, G. (2007). Evidence that long-term potentiation occurs within individual hippocampal synapses during learning. J Neurosci 27, 8031-8039.

Fernandez, M.P., Berni, J., and Ceriani, M.F. (2008). Circadian remodeling of neuronal circuits involved in rhythmic behavior. PLoS Biol 6, e69.

Fiala, A., Spall, T., Diegelmann, S., Eisermann, B., Sachse, S., Devaud, J.M., Buchner, E., and Galizia, C.G. (2002a). Genetically expressed cameleon in Drosophila melanogaster is used to visualize olfactory information in projection neurons. Curr Biol 12, 1877-1884.

Fiala, C.F., and Harris, K.M. (2001). Dendrite Structure. In Dendrites, G. Stuart, N. Spruston, and M. Hausser, eds. (Oxford, Oxford University Press), pp. 1-28.

Fiala, J.C., Spacek, J., and Harris, K.M. (2002b). Dendritic spine pathology: cause or consequence of neurological disorders? Brain Res Brain Res Rev 39, 29-54.

Fifkova, E., and Delay, R.J. (1982). Cytoplasmic actin in neuronal processes as a possible mediator of synaptic plasticity. J Cell Biol 95, 345-350.

Fischer, M., Kaech, S., Knutti, D., and Matus, A. (1998). Rapid actin-based plasticity in dendritic spines. Neuron 20, 847-854.

Fishilevich, E., and Vosshall, L.B. (2005). Genetic and functional subdivision of the Drosophila antennal lobe. Curr Biol 15, 1548-1553.

Flavell, S.W., Cowan, C.W., Kim, T.K., Greer, P.L., Lin, Y., Paradis, S., Griffith, E.C., Hu, L.S., Chen, C., and Greenberg, M.E. (2006). Activity-dependent regulation of MEF2 transcription factors suppresses excitatory synapse number. Science *311*, 1008-1012.

Frambach, I., Rossler, W., Winkler, M., and Schurmann, F.W. (2004). F-actin at identified synapses in the mushroom body neuropil of the insect brain. J Comp Neurol *475*, 303-314.

Frye, M.A., and Dickinson, M.H. (2001). Fly flight: a model for the neural control of complex behavior. Neuron *32*, 385-388.

Ganeshina, O., and Menzel, R. (2001). GABA-immunoreactive neurons in the mushroom bodies of the honeybee: an electron microscopic study. J Comp Neurol *437*, 335-349.

Ganeshina, O., Vorobyev, M., and Menzel, R. (2006). Synaptogenesis in the mushroom body calyx during metamorphosis in the honeybee Apis mellifera: an electron microscopic study. J Comp Neurol *497*, 876-897.

Gao, F.B., and Bogert, B.A. (2003). Genetic control of dendritic morphogenesis in Drosophila. Trends Neurosci *26*, 262-268.

Gao, F.B., Kohwi, M., Brenman, J.E., Jan, L.Y., and Jan, Y.N. (2000). Control of dendritic field formation in Drosophila: the roles of flamingo and competition between homologous neurons. Neuron *28*, 91-101.

Gegear, R.J., Casselman, A., Waddell, S., and Reppert, S.M. (2008). Cryptochrome mediates light-dependent magnetosensitivity in Drosophila. Nature *454*, 1014-1018.

Gerber, B., Tanimoto, H., and Heisenberg, M. (2004). An engram found? Evaluating the evidence from fruit flies. Curr Opin Neurobiol *14*, 737-744.

Ghysen, A. (2003). Dendritic arbors: a tale of living tiles. Curr Biol *13*, R427-429.

Globus, A., Rosenzweig, M.R., Bennett, E.L., and Diamond, M.C. (1973). Effects of differential experience on dendritic spine counts in rat cerebral cortex. J Comp Physiol Psychol *82*, 175-181.

Godenschwege, T.A., Simpson, J.H., Shan, X., Bashaw, G.J., Goodman, C.S., and Murphey, R.K. (2002). Ectopic expression in the giant fiber system of Drosophila reveals distinct roles for roundabout (Robo), Robo2, and Robo3 in dendritic guidance and synaptic connectivity. J Neurosci *22*, 3117-3129.

Gorska-Andrzejak, J., Keller, A., Raabe, T., Kilianek, L., and Pyza, E. (2005). Structural daily rhythms in GFP-labelled neurons in the visual system of Drosophila melanogaster. Photochem Photobiol Sci *4*, 721-726.

Govek, E.E., Newey, S.E., and Van Aelst, L. (2005). The role of the Rho GTPases in neuronal development. Genes Dev *19*, 1-49.

Grauso, M., Reenan, R.A., Culetto, E., and Sattelle, D.B. (2002). Novel putative nicotinic acetylcholine receptor subunit genes, Dalpha5, Dalpha6 and Dalpha7, in Drosophila melanogaster identify a new and highly conserved target of adenosine deaminase acting on RNA-mediated A-to-I pre-mRNA editing. Genetics *160*, 1519-1533.

Gray, E.G. (1959). Axo-somatic and axo-dendritic synapses of the cerebral cortex: an electron microscope study. J Anat *93*, 420-433.

Grieder, N.C., de Cuevas, M., and Spradling, A.C. (2000). The fusome organizes the microtubule network during oocyte differentiation in Drosophila. Development *127*, 4253-4264.

Griffith, L.C., and Budnik, V. (2006). Plasticity and second messengers during synapse development. Int Rev Neurobiol *75*, 237-265.

Groh, C., Ahrens, D., and Rossler, W. (2006). Environment- and age-dependent plasticity of synaptic complexes in the mushroom bodies of honeybee queens. Brain Behav Evol *68*, 1-14.

Groh, C., Tautz, J., and Rossler, W. (2004). Synaptic organization in the adult honey bee brain is influenced by brood-temperature control during pupal development. Proc Natl Acad Sci U S A *101*, 4268-4273.

Gronenberg, W. (2001). Subdivisions of hymenopteran mushroom body calyces by their afferent supply. J Comp Neurol *435*, 474-489.

Grueber, W.B., and Jan, Y.N. (2004). Dendritic development: lessons from Drosophila and related branches. Curr Opin Neurobiol *14*, 74-82.

Grueber, W.B., Yang, C.H., Ye, B., and Jan, Y.N. (2005). The development of neuronal morphology in insects. Curr Biol *15*, R730-738.

Grueber, W.B., Ye, B., Yang, C.H., Younger, S., Borden, K., Jan, L.Y., and Jan, Y.N. (2007). Projections of Drosophila multidendritic neurons in the central nervous system: links with peripheral dendrite morphology. Development *134*, 55-64.

Grunewald, B. (1999). Morphology of feedback neurons in the mushroom body of the honeybee, Apis mellifera. J Comp Neurol *404*, 114-126.

Guan, Z., Saraswati, S., Adolfsen, B., and Littleton, J.T. (2005). Genome-wide transcriptional changes associated with enhanced activity in the Drosophila nervous system. Neuron *48*, 91-107.

Haag, J., and Borst, A. (2002). Dendro-dendritic interactions between motion-sensitive large-field neurons in the fly. J Neurosci *22*, 3227-3233.

Haag, J., and Borst, A. (2003). Orientation tuning of motion-sensitive neurons shaped by vertical-horizontal network interactions. J Comp Physiol A Neuroethol Sens Neural Behav Physiol *189*, 363-370.

Haag, J., Theunissen, F., and Borst, A. (1997). The intrinsic electrophysiological characteristics of fly lobula plate tangential cells: II. Active membrane properties. J Comput Neurosci *4*, 349-369.

Haag, J., Vermeulen, A., and Borst, A. (1999). The intrinsic electrophysiological characteristics of fly lobula plate tangential cells: III. Visual response properties. J Comput Neurosci 7, 213-234.

Hallem, E.A., and Carlson, J.R. (2004). The odor coding system of Drosophila. Trends Genet 20, 453-459.

Hallem, E.A., Dahanukar, A., and Carlson, J.R. (2006). Insect odor and taste receptors. Annu Rev Entomol 51, 113-135.

Hallem, E.A., Ho, M.G., and Carlson, J.R. (2004). The molecular basis of odor coding in the Drosophila antenna. Cell 117, 965-979.

Harris, K.M., Jensen, F.E., and Tsao, B. (1992). Three-dimensional structure of dendritic spines and synapses in rat hippocampus (CA1) at postnatal day 15 and adult ages: implications for the maturation of synaptic physiology and long-term potentiation. J Neurosci 12, 2685-2705.

Hassan, B.A., Bermingham, N.A., He, Y., Sun, Y., Jan, Y.N., Zoghbi, H.Y., and Bellen, H.J. (2000). atonal regulates neurite arborization but does not act as a proneural gene in the Drosophila brain. Neuron 25, 549-561.

Hastings, M.H., Reddy, A.B., and Maywood, E.S. (2003). A clockwork web: circadian timing in brain and periphery, in health and disease. Nat Rev Neurosci 4, 649-661.

Hausen, K., Wolburg-Buchholz, W., and Ribi, W.A. (1980). The synaptic organization of visual interneurons in the lobula complex of flies. A light and electron microscopical study using silver-intensified cobalt-impregnations. Cell Tissue Res 208, 371-387.

Hayashi, Y., and Majewska, A.K. (2005). Dendritic spine geometry: functional implication and regulation. Neuron 46, 529-532.

Heisenberg, M. (2003). Mushroom body memoir: from maps to models. Nat Rev Neurosci 4, 266-275.

Heisenberg, M., Borst, A., Wagner, S., and Byers, D. (1985). Drosophila mushroom body mutants are deficient in olfactory learning. J Neurogenet 2, 1-30.

Heisenberg, M., Heusipp, M., and Wanke, C. (1995). Structural plasticity in the Drosophila brain. J Neurosci 15, 1951-1960.

Helfrich-Forster, C. (2005). Neurobiology of the fruit fly's circadian clock. Genes Brain Behav 4, 65-76.

Hildebrand, J.G., and Shepherd, G.M. (1997). Mechanisms of olfactory discrimination: converging evidence for common principles across phyla. Annu Rev Neurosci 20, 595-631.

Hitier, R., Heisenberg, M., and Preat, T. (1998). Abnormal mushroom body plasticity in the Drosophila memory mutant amnesiac. Neuroreport 9, 2717-2719.

Hochner, B., Shomrat, T., and Fiorito, G. (2006). The octopus: a model for a comparative analysis of the evolution of learning and memory mechanisms. The Biological bulletin 210, 308-317.

Hofer, S.B., Mrsic-Flogel, T.D., Bonhoeffer, T., and Hubener, M. (2006a). Lifelong learning: ocular dominance plasticity in mouse visual cortex. Curr Opin Neurobiol *16*, 451-459.

Hofer, S.B., Mrsic-Flogel, T.D., Bonhoeffer, T., and Hubener, M. (2006b). Prior experience enhances plasticity in adult visual cortex. Nat Neurosci *9*, 127-132.

Hofer, S.B., Mrsic-Flogel, T.D., Bonhoeffer, T., and Hubener, M. (2008). Experience leaves a lasting structural trace in cortical circuits. Nature.

Holtmaat, A., Wilbrecht, L., Knott, G.W., Welker, E., and Svoboda, K. (2006). Experience-dependent and cell-type-specific spine growth in the neocortex. Nature *441*, 979-983.

Hong, S.T., Bang, S., Hyun, S., Kang, J., Jeong, K., Paik, D., Chung, J., and Kim, J. (2008). cAMP signalling in mushroom bodies modulates temperature preference behaviour in Drosophila. Nature *454*, 771-775.

Horton, A.C., and Ehlers, M.D. (2003). Neuronal polarity and trafficking. Neuron *40*, 277-295.

Huang, B., Wang, W., Bates, M., and Zhuang, X. (2008). Three-dimensional super-resolution imaging by stochastic optical reconstruction microscopy. Science *319*, 810-813.

Huerta, R., Nowotny, T., Garcia-Sanchez, M., Abarbanel, H.D., and Rabinovich, M.I. (2004). Learning classification in the olfactory system of insects. Neural computation *16*, 1601-1640.

Isabel, G., Pascual, A., and Preat, T. (2004). Exclusive consolidated memory phases in Drosophila. Science *304*, 1024-1027.

Ismail, N., Robinson, G.E., and Fahrbach, S.E. (2006). Stimulation of muscarinic receptors mimics experience-dependent plasticity in the honey bee brain. Proc Natl Acad Sci U S A *103*, 207-211.

Ito, I., Ong, R.C., Raman, B., and Stopfer, M. (2008). Sparse odor representation and olfactory learning. Nat Neurosci *11*, 1177-1184.

Ito, K., Suzuki, K., Estes, P., Ramaswami, M., Yamamoto, D., and Strausfeld, N.J. (1998). The organization of extrinsic neurons and their implications in the functional roles of the mushroom bodies in Drosophila melanogaster Meigen. Learn Mem *5*, 52-77.

Ito, M. (2006). Cerebellar circuitry as a neuronal machine. Prog Neurobiol *78*, 272-303.

Jan, Y.N., and Jan, L.Y. (2003). The control of dendrite development. Neuron *40*, 229-242.

Jarman, A.P., Grau, Y., Jan, L.Y., and Jan, Y.N. (1993). atonal is a proneural gene that directs chordotonal organ formation in the Drosophila peripheral nervous system. Cell *73*, 1307-1321.

Jefferis, G.S., Marin, E.C., Komiyama, T., Zhu, H., Chihara, T., Berdnik, D., and Luo, L. (2005). Development of wiring specificity of the Drosophila olfactory system. Chem Senses *30 Suppl 1*, i94.

Jefferis, G.S., Marin, E.C., Stocker, R.F., and Luo, L. (2001). Target neuron prespecification in the olfactory map of Drosophila. Nature *414*, 204-208.

Jefferis, G.S., Marin, E.C., Watts, R.J., and Luo, L. (2002). Development of neuronal connectivity in Drosophila antennal lobes and mushroom bodies. Curr Opin Neurobiol *12*, 80-86.

Jefferis, G.S., Potter, C.J., Chan, A.M., Marin, E.C., Rohlfing, T., Maurer, C.R., Jr., and Luo, L. (2007). Comprehensive maps of Drosophila higher olfactory centers: spatially segregated fruit and pheromone representation. Cell *128*, 1187-1203.

Jefferis, G.S., Vyas, R.M., Berdnik, D., Ramaekers, A., Stocker, R.F., Tanaka, N.K., Ito, K., and Luo, L. (2004). Developmental origin of wiring specificity in the olfactory system of Drosophila. Development *131*, 117-130.

Jin, X., Ha, T.S., and Smith, D.P. (2008). SNMP is a signaling component required for pheromone sensitivity in Drosophila. Proc Natl Acad Sci U S A *105*, 10996-11001.

Jinushi-Nakao, S., Arvind, R., Amikura, R., Kinameri, E., Liu, A.W., and Moore, A.W. (2007). Knot/Collier and cut control different aspects of dendrite cytoskeleton and synergize to define final arbor shape. Neuron *56*, 963-978.

Joesch, M., Plett, J., Borst, A., and Reiff, D.F. (2008). Response properties of motion-sensitive visual interneurons in the lobula plate of Drosophila melanogaster. Curr Biol *18*, 368-374.

Joiner, W.J., Crocker, A., White, B.H., and Sehgal, A. (2006). Sleep in Drosophila is regulated by adult mushroom bodies. Nature *441*, 757-760.

Jorgensen, E.M., and Mango, S.E. (2002). The art and design of genetic screens: caenorhabditis elegans. Nat Rev Genet *3*, 356-369.

Jortner, R.A., Farivar, S.S., and Laurent, G. (2007). A simple connectivity scheme for sparse coding in an olfactory system. J Neurosci *27*, 1659-1669.

Kandel, E.R. (2001). The molecular biology of memory storage: a dialogue between genes and synapses. Science *294*, 1030-1038.

Kaytor, M.D., and Orr, H.T. (2001). RNA targets of the fragile X protein. Cell *107*, 555-557.

Keene, A.C., Stratmann, M., Keller, A., Perrat, P.N., Vosshall, L.B., and Waddell, S. (2004). Diverse odor-conditioned memories require uniquely timed dorsal paired medial neuron output. Neuron *44*, 521-533.

Keene, A.C., and Waddell, S. (2007). Drosophila olfactory memory: single genes to complex neural circuits. Nat Rev Neurosci *8*, 341-354.

Keleman, K., Kruttner, S., Alenius, M., and Dickson, B.J. (2007). Function of the Drosophila CPEB protein Orb2 in long-term courtship memory. Nat Neurosci *10*, 1587-1593.

Kerr, J.N., and Denk, W. (2008). Imaging in vivo: watching the brain in action. Nat Rev Neurosci *9*, 195-205.

Kim, S., and Chiba, A. (2004). Dendritic guidance. Trends Neurosci *27*, 194-202.

Kitamoto, T. (2001). Conditional modification of behavior in Drosophila by targeted expression of a temperature-sensitive shibire allele in defined neurons. J Neurobiol 47, 81-92.

Kittel, R.J., Wichmann, C., Rasse, T.M., Fouquet, W., Schmidt, M., Schmid, A., Wagh, D.A., Pawlu, C., Kellner, R.R., Willig, K.I., et al. (2006). Bruchpilot promotes active zone assembly, Ca2+ channel clustering, and vesicle release. Science 312, 1051-1054.

Kobayashi, M., Michaut, L., Ino, A., Honjo, K., Nakajima, T., Maruyama, Y., Mochizuki, H., Ando, M., Ghangrekar, I., Takahashi, K., et al. (2006). Differential microarray analysis of Drosophila mushroom body transcripts using chemical ablation. Proc Natl Acad Sci U S A 103, 14417-14422.

Komiyama, T., and Luo, L. (2006). Development of wiring specificity in the olfactory system. Curr Opin Neurobiol 16, 67-73.

Konur, S., and Yuste, R. (2004). Developmental regulation of spine and filopodial motility in primary visual cortex: reduced effects of activity and sensory deprivation. J Neurobiol 59, 236-246.

Krashes, M.J., Keene, A.C., Leung, B., Armstrong, J.D., and Waddell, S. (2007). Sequential use of mushroom body neuron subsets during drosophila odor memory processing. Neuron 53, 103-115.

Krofczik, S., Khojasteh, U., de Ibarra, N.H., and Menzel, R. (2008). Adaptation of microglomerular complexes in the honeybee mushroom body lip to manipulations of behavioral maturation and sensory experience. Dev Neurobiol 68, 1007-1017.

Kuhn-Buhlmann, S., and Wehner, R. (2006). Age-dependent and task-related volume changes in the mushroom bodies of visually guided desert ants, Cataglyphis bicolor. J Neurobiol 66, 511-521.

Kurtovic, A., Widmer, A., and Dickson, B.J. (2007). A single class of olfactory neurons mediates behavioural responses to a Drosophila sex pheromone. Nature 446, 542-546.

Lamprecht, R., and LeDoux, J. (2004). Structural plasticity and memory. Nat Rev Neurosci 5, 45-54.

Lang, S.B., Stein, V., Bonhoeffer, T., and Lohmann, C. (2007). Endogenous brain-derived neurotrophic factor triggers fast calcium transients at synapses in developing dendrites. J Neurosci 27, 1097-1105.

Larsson, M.C., Hansson, B.S., and Strausfeld, N.J. (2004). A simple mushroom body in an African scarabid beetle. J Comp Neurol 478, 219-232.

Laughlin, J.D., Ha, T.S., Jones, D.N., and Smith, D.P. (2008). Activation of pheromone-sensitive neurons is mediated by conformational activation of pheromone-binding protein. Cell 133, 1255-1265.

Laurent, G. (2002). Olfactory network dynamics and the coding of multidimensional signals. Nat Rev Neurosci 3, 884-895.

Laurent, G. (2006). Dynamics and Computation beyond the Receptor Neurons (Cambridge, MA, MIT Press books).

Laurent, G., and Naraghi, M. (1994). Odorant-induced oscillations in the mushroom bodies of the locust. J Neurosci 14, 2993-3004.

Lee, T., Lee, A., and Luo, L. (1999). Development of the Drosophila mushroom bodies: sequential generation of three distinct types of neurons from a neuroblast. Development 126, 4065-4076.

Lee, T., and Luo, L. (1999). Mosaic analysis with a repressible cell marker for studies of gene function in neuronal morphogenesis. Neuron 22, 451-461.

Lee, T., and Luo, L. (2001). Mosaic analysis with a repressible cell marker (MARCM) for Drosophila neural development. Trends Neurosci 24, 251-254.

Leiss, F., Koper, E., Hein, I., Fouquet, W., Lindner, J., Sigrist, S., and Tavosanis, G. (2009). Characterization of dendritic spines in the Drosophila central nervous system. Dev Neurobiol 69, 221-234.

Lendvai, B., Stern, E.A., Chen, B., and Svoboda, K. (2000). Experience-dependent plasticity of dendritic spines in the developing rat barrel cortex in vivo. Nature 404, 876-881.

Lent, D.D., Pinter, M., and Strausfeld, N.J. (2007). Learning with half a brain. Dev Neurobiol 67, 740-751.

Libersat, F. (2005). Maturation of dendritic architecture: lessons from insect identified neurons. J Neurobiol 64, 11-23.

Libersat, F., and Duch, C. (2004). Mechanisms of dendritic maturation. Mol Neurobiol 29, 303-320.

Lin, H.H., Lai, J.S., Chin, A.L., Chen, Y.C., and Chiang, A.S. (2007). A map of olfactory representation in the Drosophila mushroom body. Cell 128, 1205-1217.

Lippman, J., and Dunaevsky, A. (2005). Dendritic spine morphogenesis and plasticity. J Neurobiol 64, 47-57.

Lisman, J., Schulman, H., and Cline, H. (2002). The molecular basis of CaMKII function in synaptic and behavioural memory. Nat Rev Neurosci 3, 175-190.

Littleton, J.T., Bellen, H.J., and Perin, M.S. (1993). Expression of synaptotagmin in Drosophila reveals transport and localization of synaptic vesicles to the synapse. Development 118, 1077-1088.

Liu, X., Krause, W.C., and Davis, R.L. (2007). GABAA receptor RDL inhibits Drosophila olfactory associative learning. Neuron 56, 1090-1102.

London, M., and Hausser, M. (2005). Dendritic computation. Annu Rev Neurosci 28, 503-532.

Losonczy, A., Makara, J.K., and Magee, J.C. (2008). Compartmentalized dendritic plasticity and input feature storage in neurons. Nature 452, 436-441.

Luo, L., Hensch, T.K., Ackerman, L., Barbel, S., Jan, L.Y., and Jan, Y.N. (1996). Differential effects of the Rac GTPase on Purkinje cell axons and dendritic trunks and spines. Nature *379*, 837-840.

Luo, L., Liao, Y.J., Jan, L.Y., and Jan, Y.N. (1994). Distinct morphogenetic functions of similar small GTPases: Drosophila Drac1 is involved in axonal outgrowth and myoblast fusion. Genes Dev *8*, 1787-1802.

Lyons, G.E., Micales, B.K., Schwarz, J., Martin, J.F., and Olson, E.N. (1995). Expression of mef2 genes in the mouse central nervous system suggests a role in neuronal maturation. J Neurosci *15*, 5727-5738.

Majewska, A., and Sur, M. (2003). Motility of dendritic spines in visual cortex in vivo: changes during the critical period and effects of visual deprivation. Proc Natl Acad Sci U S A *100*, 16024-16029.

Maletic-Savatic, M., Malinow, R., and Svoboda, K. (1999). Rapid dendritic morphogenesis in CA1 hippocampal dendrites induced by synaptic activity. Science *283*, 1923-1927.

Marek, K.W., and Davis, G.W. (2003). Controlling the active properties of excitable cells. Curr Opin Neurobiol *13*, 607-611.

Mares, S., Ash, L., and Gronenberg, W. (2005). Brain allometry in bumblebee and honey bee workers. Brain Behav Evol *66*, 50-61.

Margulies, C., Tully, T., and Dubnau, J. (2005). Deconstructing memory in Drosophila. Curr Biol *15*, R700-713.

Marin, E.C., Jefferis, G.S., Komiyama, T., Zhu, H., and Luo, L. (2002). Representation of the glomerular olfactory map in the Drosophila brain. Cell *109*, 243-255.

Masuda-Nakagawa, L.M., Tanaka, N.K., and O'Kane, C.J. (2005). Stereotypic and random patterns of connectivity in the larval mushroom body calyx of Drosophila. Proc Natl Acad Sci U S A *102*, 19027-19032.

Matus, A. (2000). Actin-based plasticity in dendritic spines. Science *290*, 754-758.

Matus, A. (2005). Growth of dendritic spines: a continuing story. Curr Opin Neurobiol *15*, 67-72.

Matus, A., Ackermann, M., Pehling, G., Byers, H.R., and Fujiwara, K. (1982). High actin concentrations in brain dendritic spines and postsynaptic densities. Proc Natl Acad Sci U S A *79*, 7590-7594.

Mazzoni, E.O., Desplan, C., and Blau, J. (2005). Circadian pacemaker neurons transmit and modulate visual information to control a rapid behavioral response. Neuron *45*, 293-300.

McGuire, S.E., Le, P.T., and Davis, R.L. (2001). The role of Drosophila mushroom body signaling in olfactory memory. Science *293*, 1330-1333.

McGuire, S.E., Le, P.T., Osborn, A.J., Matsumoto, K., and Davis, R.L. (2003). Spatiotemporal rescue of memory dysfunction in Drosophila. Science *302*, 1765-1768.

Medina, P.M., Worthen, R.J., Forsberg, L.J., and Brenman, J.E. (2008). The actin-binding protein capulet genetically interacts with the microtubule motor kinesin to maintain neuronal dendrite homeostasis. PLoS ONE 3, e3054.

Mehren, J.E., and Griffith, L.C. (2006). Cholinergic neurons mediate CaMKII-dependent enhancement of courtship suppression. Learn Mem 13, 686-689.

Micheva, K.D., and Smith, S.J. (2007). Array tomography: a new tool for imaging the molecular architecture and ultrastructure of neural circuits. Neuron 55, 25-36.

Mizrahi, A., and Libersat, F. (2002). Afferent input regulates the formation of distal dendritic branches. J Comp Neurol 452, 1-10.

Mobbs, P.G. (1982). The Brain of the Honeybee Apis Mellifera. I. The Connections and Spatial Organization of the Mushroom Bodies. Philos Trans R Soc Lond, 309-354.

Moore, A.W., Jan, L.Y., and Jan, Y.N. (2002). hamlet, a binary genetic switch between single- and multiple- dendrite neuron morphology. Science 297, 1355-1358.

Murthy, M., Fiete, I., and Laurent, G. (2008). Testing odor response stereotypy in the Drosophila mushroom body. Neuron 59, 1009-1023.

Nagerl, U.V., Eberhorn, N., Cambridge, S.B., and Bonhoeffer, T. (2004). Bidirectional activity-dependent morphological plasticity in hippocampal neurons. Neuron 44, 759-767.

Nakayama, A.Y., Harms, M.B., and Luo, L. (2000). Small GTPases Rac and Rho in the maintenance of dendritic spines and branches in hippocampal pyramidal neurons. J Neurosci 20, 5329-5338.

Newsome, T.P., Asling, B., and Dickson, B.J. (2000). Analysis of Drosophila photoreceptor axon guidance in eye-specific mosaics. Development 127, 851-860.

Ng, M., Roorda, R.D., Lima, S.Q., Zemelman, B.V., Morcillo, P., and Miesenbock, G. (2002). Transmission of olfactory information between three populations of neurons in the antennal lobe of the fly. Neuron 36, 463-474.

Nieus, T., Sola, E., Mapelli, J., Saftenku, E., Rossi, P., and D'Angelo, E. (2006). LTP regulates burst initiation and frequency at mossy fiber-granule cell synapses of rat cerebellum: experimental observations and theoretical predictions. J Neurophysiol 95, 686-699.

Nitabach, M.N., Blau, J., and Holmes, T.C. (2002). Electrical silencing of Drosophila pacemaker neurons stops the free-running circadian clock. Cell 109, 485-495.

Noguchi, J., Matsuzaki, M., Ellis-Davies, G.C., and Kasai, H. (2005). Spine-neck geometry determines NMDA receptor-dependent Ca2+ signaling in dendrites. Neuron 46, 609-622.

Nowotny, T., and Huerta, R. (2003). Explaining synchrony in feed-forward networks: are McCulloch-Pitts neurons good enough? Biol Cybern 89, 237-241.

Nowotny, T., Huerta, R., Abarbanel, H.D., and Rabinovich, M.I. (2005). Self-organization in the olfactory system: one shot odor recognition in insects. Biol Cybern 93, 436-446.

Nusslein-Volhard, C., and Wieschaus, E. (1980). Mutations affecting segment number and polarity in Drosophila. Nature 287, 795-801.

Olsen, S.R., and Wilson, R.I. (2008). Lateral presynaptic inhibition mediates gain control in an olfactory circuit. Nature 452, 956-960.

Oray, S., Majewska, A., and Sur, M. (2004). Dendritic spine dynamics are regulated by monocular deprivation and extracellular matrix degradation. Neuron 44, 1021-1030.

Oray, S., Majewska, A., and Sur, M. (2006). Effects of synaptic activity on dendritic spine motility of developing cortical layer v pyramidal neurons. Cereb Cortex 16, 730-741.

Page, D.R., and Grossniklaus, U. (2002). The art and design of genetic screens: Arabidopsis thaliana. Nat Rev Genet 3, 124-136.

Pan, L., Zhang, Y.Q., Woodruff, E., and Broadie, K. (2004). The Drosophila fragile X gene negatively regulates neuronal elaboration and synaptic differentiation. Curr Biol 14, 1863-1870.

Parrish, J.Z., Emoto, K., Kim, M.D., and Jan, Y.N. (2007). Mechanisms that regulate establishment, maintenance, and remodeling of dendritic fields. Annu Rev Neurosci 30, 399-423.

Pascual, A., and Preat, T. (2001). Localization of long-term memory within the Drosophila mushroom body. Science 294, 1115-1117.

Perez Orive, J., Mazor, O., Turner, G.C., Cassenaer, S., Wilson, R.I., and Laurent, G. (2002). Oscillations and sparsening of odor representations in the mushroom body. Science 297, 359-365.

Phelan, P., Nakagawa, M., Wilkin, M.B., Moffat, K.G., O'Kane, C.J., Davies, J.A., and Bacon, J.P. (1996). Mutations in shaking-B prevent electrical synapse formation in the Drosophila giant fiber system. J Neurosci 16, 1101-1113.

Pierantoni, R. (1976). A look into the cock-pit of the fly. The architecture of the lobular plate. Cell Tissue Res 171, 101-122.

Pitman, J.L., McGill, J.J., Keegan, K.P., and Allada, R. (2006). A dynamic role for the mushroom bodies in promoting sleep in Drosophila. Nature 441, 753-756.

Prokop, A., and Meinertzhagen, I.A. (2006). Development and structure of synaptic contacts in Drosophila. Semin Cell Dev Biol 17, 20-30.

Raghu, S.V., Joesch, M., Borst, A., and Reiff, D.F. (2007). Synaptic organization of lobula plate tangential cells in Drosophila: gamma-aminobutyric acid receptors and chemical release sites. J Comp Neurol 502, 598-610.

Ramaekers, A., Magnenat, E., Marin, E.C., Gendre, N., Jefferis, G.S., Luo, L., and Stocker, R.F. (2005). Glomerular maps without cellular redundancy at successive levels of the Drosophila larval olfactory circuit. Curr Biol 15, 982-992.

Ramnani, N. (2006). The primate cortico-cerebellar system: anatomy and function. Nat Rev Neurosci 7, 511-522.

Renn, S.C., Park, J.H., Rosbash, M., Hall, J.C., and Taghert, P.H. (1999). A pdf neuropeptide gene mutation and ablation of PDF neurons each cause severe abnormalities of behavioral circadian rhythms in Drosophila. Cell *99*, 791-802.

Reuter, J.E., Nardine, T.M., Penton, A., Billuart, P., Scott, E.K., Usui, T., Uemura, T., and Luo, L. (2003). A mosaic genetic screen for genes necessary for Drosophila mushroom body neuronal morphogenesis. Development *130*, 1203-1213.

Sachse, S., Rueckert, E., Keller, A., Okada, R., Tanaka, N.K., Ito, K., and Vosshall, L.B. (2007). Activity-dependent plasticity in an olfactory circuit. Neuron *56*, 838-850.

Sanchez-Soriano, N., Bottenberg, W., Fiala, A., Haessler, U., Kerassoviti, A., Knust, E., Lohr, R., and Prokop, A. (2005). Are dendrites in Drosophila homologous to vertebrate dendrites? Dev Biol *288*, 126-138.

Schmitt, S., Evers, J.F., Duch, C., Scholz, M., and Obermayer, K. (2004). New methods for the computer-assisted 3-D reconstruction of neurons from confocal image stacks. NeuroImage *23*, 1283-1298.

Schnorrer, F., and Dickson, B.J. (2004). Axon guidance: morphogens show the way. Curr Biol *14*, R19-21.

Schubert, V., and Dotti, C.G. (2007). Transmitting on actin: synaptic control of dendritic architecture. J Cell Sci *120*, 205-212.

Schurmann, F.W. (1974). [On the functional anatomy of the corpora pedunculata in insects (author's transl)]. Exp Brain Res *19*, 406-432.

Schurmann, F.W., Ottersen, O.P., and Honegger, H.W. (2000). Glutamate-like immunoreactivity marks compartments of the mushroom bodies in the brain of the cricket. J Comp Neurol *418*, 227-239.

Scott, E.K., Raabe, T., and Luo, L. (2002). Structure of the vertical and horizontal system neurons of the lobula plate in Drosophila. J Comp Neurol *454*, 470-481.

Scott, E.K., Reuter, J.E., and Luo, L. (2003a). Dendritic development of Drosophila high order visual system neurons is independent of sensory experience. BMC Neurosci *4*, 14.

Scott, E.K., Reuter, J.E., and Luo, L. (2003b). Small GTPase Cdc42 is required for multiple aspects of dendritic morphogenesis. J Neurosci *23*, 3118-3123.

Segal, M. (2005). Dendritic spines and long-term plasticity. Nat Rev Neurosci *6*, 277-284.

Seid, M.A., Harris, K.M., and Traniello, J.F. (2005). Age-related changes in the number and structure of synapses in the lip region of the mushroom bodies in the ant Pheidole dentata. J Comp Neurol *488*, 269-277.

Sepp, K.J., Schulte, J., and Auld, V.J. (2001). Peripheral glia direct axon guidance across the CNS/PNS transition zone. Dev Biol *238*, 47-63.

Shalizi, A., Gaudilliere, B., Yuan, Z., Stegmuller, J., Shirogane, T., Ge, Q., Tan, Y., Schulman, B., Harper, J.W., and Bonni, A. (2006). A calcium-regulated MEF2 sumoylation switch controls postsynaptic differentiation. Science *311*, 1012-1017.

Shang, Y., Claridge-Chang, A., Sjulson, L., Pypaert, M., and Miesenbock, G. (2007). Excitatory local circuits and their implications for olfactory processing in the fly antennal lobe. Cell *128*, 601-612.

Shen, H., Sesack, S.R., Toda, S., and Kalivas, P.W. (2008). Automated quantification of dendritic spine density and spine head diameter in medium spiny neurons of the nucleus accumbens. Brain structure & function *213*, 149-157.

Shomrat, T., Zarrella, I., Fiorito, G., and Hochner, B. (2008). The octopus vertical lobe modulates short-term learning rate and uses LTP to acquire long-term memory. Curr Biol *18*, 337-342.

Sigg, D., Thompson, C.M., and Mercer, A.R. (1997). Activity-dependent changes to the brain and behavior of the honey bee, Apis mellifera (L.). J Neurosci *17*, 7148-7156.

Sinakevitch, I., Farris, S.M., and Strausfeld, N.J. (2001). Taurine-, aspartate- and glutamate-like immunoreactivity identifies chemically distinct subdivisions of Kenyon cells in the cockroach mushroom body. J Comp Neurol *439*, 352-367.

Single, S., and Borst, A. (1998). Dendritic integration and its role in computing image velocity. Science *281*, 1848-1850.

Sjoholm, M., Sinakevitch, I., Ignell, R., Strausfeld, N.J., and Hansson, B.S. (2005). Organization of Kenyon cells in subdivisions of the mushroom bodies of a lepidopteran insect. J Comp Neurol *491*, 290-304.

Smith, D., Wessnitzer, J., and Webb, B. (2008). A model of associative learning in the mushroom body. Biol Cybern *99*, 89-103.

Spradling, A.C., Stern, D., Beaton, A., Rhem, E.J., Laverty, T., Mozden, N., Misra, S., and Rubin, G.M. (1999). The Berkeley Drosophila Genome Project gene disruption project: Single P-element insertions mutating 25% of vital Drosophila genes. Genetics *153*, 135-177.

St Johnston, D. (2002). The art and design of genetic screens: Drosophila melanogaster. Nat Rev Genet *3*, 176-188.

Stanewsky, R. (2003). Genetic analysis of the circadian system in Drosophila melanogaster and mammals. J Neurobiol *54*, 111-147.

Stocker, R.F. (1994). The organization of the chemosensory system in Drosophila melanogaster: a review. Cell Tissue Res *275*, 3-26.

Stocker, R.F., Heimbeck, G., Gendre, N., and de Belle, J.S. (1997). Neuroblast ablation in Drosophila P[GAL4] lines reveals origins of olfactory interneurons. J Neurobiol *32*, 443-456.

Stocker, R.F., Lienhard, M.C., Borst, A., and Fischbach, K.F. (1990). Neuronal architecture of the antennal lobe in Drosophila melanogaster. Cell Tissue Res *262*, 9-34.

Stopfer, M. (2005). Olfactory coding: inhibition reshapes odor responses. Curr Biol 15, R996-998.

Stopfer, M., Jayaraman, V., and Laurent, G. (2003). Intensity versus identity coding in an olfactory system. Neuron 39, 991-1004.

Strausfeld, N.J. (2002). Organization of the honey bee mushroom body: representation of the calyx within the vertical and gamma lobes. J Comp Neurol 450, 4-33.

Strausfeld, N.J., Hansen, L., Li, Y., Gomez, R.S., and Ito, K. (1998). Evolution, discovery, and interpretations of arthropod mushroom bodies. Learn Mem 5, 11-37.

Strausfeld, N.J., and Lee, J.K. (1991). Neuronal basis for parallel visual processing in the fly. Vis Neurosci 7, 13-33.

Strausfeld, N.J., Sinakevitch, I., and Vilinsky, I. (2003). The mushroom bodies of Drosophila melanogaster: an immunocytological and golgi study of Kenyon cell organization in the calyces and lobes. Microsc Res Tech 62, 151-169.

Sutton, M.A., and Schuman, E.M. (2006). Dendritic protein synthesis, synaptic plasticity, and memory. Cell 127, 49-58.

Szyszka, P., Ditzen, M., Galkin, A., Galizia, C.G., and Menzel, R. (2005). Sparsening and temporal sharpening of olfactory representations in the honeybee mushroom bodies. J Neurophysiol 94, 3303-3313.

Tada, T., and Sheng, M. (2006). Molecular mechanisms of dendritic spine morphogenesis. Curr Opin Neurobiol 16, 95-101.

Tanaka, N.K., Awasaki, T., Shimada, T., and Ito, K. (2004). Integration of chemosensory pathways in the Drosophila second-order olfactory centers. Curr Biol 14, 449-457.

Tashiro, A., and Yuste, R. (2004). Regulation of dendritic spine motility and stability by Rac1 and Rho kinase: evidence for two forms of spine motility. Mol Cell Neurosci 26, 429-440.

Tessier, C.R., and Broadie, K. (2008). Drosophila fragile X mental retardation protein developmentally regulates activity-dependent axon pruning. Development 135, 1547-1557.

Thum, A.S., Knapek, S., Rister, J., Dierichs-Schmitt, E., Heisenberg, M., and Tanimoto, H. (2006). Differential potencies of effector genes in adult Drosophila. J Comp Neurol 498, 194-203.

Trachtenberg, J.T., Chen, B.E., Knott, G.W., Feng, G., Sanes, J.R., Welker, E., and Svoboda, K. (2002). Long-term in vivo imaging of experience-dependent synaptic plasticity in adult cortex. Nature 420, 788-794.

Trimarchi, J.R., Jin, P., and Murphey, R.K. (1999). Controlling the motor neuron. Int Rev Neurobiol 43, 241-264.

Turner, G.C., Bazhenov, M., and Laurent, G. (2008). Olfactory representations by Drosophila mushroom body neurons. J Neurophysiol 99, 734-746.

Venken, K.J., and Bellen, H.J. (2005). Emerging technologies for gene manipulation in Drosophila melanogaster. Nat Rev Genet *6*, 167-178.

Verkhusha, V.V., Tsukita, S., and Oda, H. (1999). Actin dynamics in lamellipodia of migrating border cells in the Drosophila ovary revealed by a GFP-actin fusion protein. FEBS Lett *445*, 395-401.

Vosshall, L.B., Amrein, H., Morozov, P.S., Rzhetsky, A., and Axel, R. (1999). A spatial map of olfactory receptor expression in the Drosophila antenna. Cell *96*, 725-736.

Waddell, S., Armstrong, J.D., Kitamoto, T., Kaiser, K., and Quinn, W.G. (2000). The amnesiac gene product is expressed in two neurons in the Drosophila brain that are critical for memory. Cell *103*, 805-813.

Wagh, D.A., Rasse, T.M., Asan, E., Hofbauer, A., Schwenkert, I., Durrbeck, H., Buchner, S., Dabauvalle, M.C., Schmidt, M., Qin, G., *et al.* (2006). Bruchpilot, a protein with homology to ELKS/CAST, is required for structural integrity and function of synaptic active zones in Drosophila. Neuron *49*, 833-844.

Wall, M.J. (2005). Short-term synaptic plasticity during development of rat mossy fibre to granule cell synapses. Eur J Neurosci *21*, 2149-2158.

Wang, J.W., Wong, A.M., Flores, J., Vosshall, L.B., and Axel, R. (2003). Two-photon calcium imaging reveals an odor-evoked map of activity in the fly brain. Cell *112*, 271-282.

Wang, L., and Zheng, Y. (2007). Cell type-specific functions of Rho GTPases revealed by gene targeting in mice. Trends Cell Biol *17*, 58-64.

Wang, V.Y., Hassan, B.A., Bellen, H.J., and Zoghbi, H.Y. (2002). Drosophila atonal fully rescues the phenotype of Math1 null mice: new functions evolve in new cellular contexts. Curr Biol *12*, 1611-1616.

Weeks, J.C., and Truman, J.W. (1985). Independent steroid control of the fates of motoneurons and their muscles during insect metamorphosis. J Neurosci *5*, 2290-2300.

Wilson, R.I., and Laurent, G. (2005). Role of GABAergic inhibition in shaping odor-evoked spatiotemporal patterns in the Drosophila antennal lobe. J Neurosci *25*, 9069-9079.

Wilson, R.I., and Mainen, Z.F. (2006). Early events in olfactory processing. Annu Rev Neurosci *29*, 163-201.

Wilson, R.I., Turner, G.C., and Laurent, G. (2004). Transformation of olfactory representations in the Drosophila antennal lobe. Science *303*, 366-370.

Withers, G.S., Day, N.F., Talbot, E.F., Dobson, H.E., and Wallace, C.S. (2008). Experience-dependent plasticity in the mushroom bodies of the solitary bee Osmia lignaria (Megachilidae). Dev Neurobiol *68*, 73-82.

Withers, G.S., Fahrbach, S.E., and Robinson, G.E. (1993). Selective neuroanatomical plasticity and division of labour in the honeybee. Nature *364*, 238-240.

Wong, A.M., Wang, J.W., and Axel, R. (2002). Spatial representation of the glomerular map in the Drosophila protocerebrum. Cell *109*, 229-241.

Wong, R.O., and Ghosh, A. (2002). Activity-dependent regulation of dendritic growth and patterning. Nat Rev Neurosci *3*, 803-812.

Wu, J.S., and Luo, L. (2006). A protocol for dissecting Drosophila melanogaster brains for live imaging or immunostaining. Nat Protoc *1*, 2110-2115.

Xu-Friedman, M.A., and Regehr, W.G. (2003). Ultrastructural contributions to desensitization at cerebellar mossy fiber to granule cell synapses. J Neurosci *23*, 2182-2192.

Xu, P., Atkinson, R., Jones, D.N., and Smith, D.P. (2005). Drosophila OBP LUSH is required for activity of pheromone-sensitive neurons. Neuron *45*, 193-200.

Yamazaki, Y., Nishikawa, M., and Mizunami, M. (1998). Three classes of GABA-like immunoreactive neurons in the mushroom body of the cockroach. Brain Res *788*, 80-86.

Yang, M.Y., Armstrong, J.D., Vilinsky, I., Strausfeld, N.J., and Kaiser, K. (1995). Subdivision of the Drosophila mushroom bodies by enhancer-trap expression patterns. Neuron *15*, 45-54.

Yasuyama, K., Meinertzhagen, I.A., and Schurmann, F.W. (2002). Synaptic organization of the mushroom body calyx in Drosophila melanogaster. J Comp Neurol *445*, 211-226.

Yasuyama, K., Meinertzhagen, I.A., and Schurmann, F.W. (2003). Synaptic connections of cholinergic antennal lobe relay neurons innervating the lateral horn neuropile in the brain of Drosophila melanogaster. J Comp Neurol *466*, 299-315.

Yuste, R., and Bonhoeffer, T. (2001). Morphological changes in dendritic spines associated with long-term synaptic plasticity. Annu Rev Neurosci *24*, 1071-1089.

Yuste, R., and Bonhoeffer, T. (2004). Genesis of dendritic spines: insights from ultrastructural and imaging studies. Nat Rev Neurosci *5*, 24-34.

Yuste, R., Majewska, A., and Holthoff, K. (2000). From form to function: calcium compartmentalization in dendritic spines. Nat Neurosci *3*, 653-659.

Zars, T., Fischer, M., Schulz, R., and Heisenberg, M. (2000). Localization of a short-term memory in Drosophila. Science *288*, 672-675.

Zhai, R.G., and Bellen, H.J. (2004). The architecture of the active zone in the presynaptic nerve terminal. Physiology (Bethesda) *19*, 262-270.

Zhang, F., Wang, L.P., Brauner, M., Liewald, J.F., Kay, K., Watzke, N., Wood, P.G., Bamberg, E., Nagel, G., Gottschalk, A., *et al.* (2007). Multimodal fast optical interrogation of neural circuitry. Nature *446*, 633-639.

Zhang, H., and Macara, I.G. (2006). The polarity protein PAR-3 and TIAM1 cooperate in dendritic spine morphogenesis. Nat Cell Biol *8*, 227-237.

Zhang, Y.Q., Bailey, A.M., Matthies, H.J., Renden, R.B., Smith, M.A., Speese, S.D., Rubin, G.M., and Broadie, K. (2001). Drosophila fragile X-related gene regulates the MAP1B homolog Futsch to control synaptic structure and function. Cell *107*, 591-603.

Zhu, S., Chiang, A.S., and Lee, T. (2003). Development of the Drosophila mushroom bodies: elaboration, remodeling and spatial organization of dendrites in the calyx. Development *130*, 2603-2610.

Zhu, S., Perez, R., Pan, M., and Lee, T. (2005). Requirement of Cul3 for axonal arborization and dendritic elaboration in Drosophila mushroom body neurons. J Neurosci *25*, 4189-4197.

Zou, Z., Horowitz, L.F., Montmayeur, J.P., Snapper, S., and Buck, L.B. (2001). Genetic tracing reveals a stereotyped sensory map in the olfactory cortex. Nature *414*, 173-179.

Die VDM Verlagsservicegesellschaft sucht für wissenschaftliche Verlage abgeschlossene und herausragende

Dissertationen, Habilitationen, Diplomarbeiten, Master Theses, Magisterarbeiten usw.

für die kostenlose Publikation als Fachbuch.

Sie verfügen über eine Arbeit, die hohen inhaltlichen und formalen Ansprüchen genügt, und haben Interesse an einer honorarvergüteten Publikation?

Dann senden Sie bitte erste Informationen über sich und Ihre Arbeit per Email an *info@vdm-vsg.de*.

Sie erhalten kurzfristig unser Feedback!

VDM Verlagsservicegesellschaft mbH
Dudweiler Landstr. 99
D - 66123 Saarbrücken

Telefon +49 681 3720 174
Fax +49 681 3720 1749

www.vdm-vsg.de

Die VDM Verlagsservicegesellschaft mbH vertritt

Printed by Books on Demand GmbH, Norderstedt / Germany